Carsten Enderlein

The Interaction of Graphene with Different Substrates

Carsten Enderlein

The Interaction of Graphene with Different Substrates

An ARPES study

Südwestdeutscher Verlag für Hochschulschriften

Impressum / Imprint
Bibliografische Information der Deutschen Nationalbibliothek: Die Deutsche Nationalbibliothek verzeichnet diese Publikation in der Deutschen Nationalbibliografie; detaillierte bibliografische Daten sind im Internet über http://dnb.d-nb.de abrufbar.
Alle in diesem Buch genannten Marken und Produktnamen unterliegen warenzeichen-, marken- oder patentrechtlichem Schutz bzw. sind Warenzeichen oder eingetragene Warenzeichen der jeweiligen Inhaber. Die Wiedergabe von Marken, Produktnamen, Gebrauchsnamen, Handelsnamen, Warenbezeichnungen u.s.w. in diesem Werk berechtigt auch ohne besondere Kennzeichnung nicht zu der Annahme, dass solche Namen im Sinne der Warenzeichen- und Markenschutzgesetzgebung als frei zu betrachten wären und daher von jedermann benutzt werden dürften.

Bibliographic information published by the Deutsche Nationalbibliothek: The Deutsche Nationalbibliothek lists this publication in the Deutsche Nationalbibliografie; detailed bibliographic data are available in the Internet at http://dnb.d-nb.de.
Any brand names and product names mentioned in this book are subject to trademark, brand or patent protection and are trademarks or registered trademarks of their respective holders. The use of brand names, product names, common names, trade names, product descriptions etc. even without a particular marking in this works is in no way to be construed to mean that such names may be regarded as unrestricted in respect of trademark and brand protection legislation and could thus be used by anyone.

Coverbild / Cover image: www.ingimage.com

Verlag / Publisher:
Südwestdeutscher Verlag für Hochschulschriften
ist ein Imprint der / is a trademark of
AV Akademikerverlag GmbH & Co. KG
Heinrich-Böcking-Str. 6-8, 66121 Saarbrücken, Deutschland / Germany
Email: info@svh-verlag.de

Herstellung: siehe letzte Seite /
Printed at: see last page
ISBN: 978-3-8381-3557-1

Zugl. / Approved by: Berlin, Freie Universität, Dissertation, 2010

Copyright © 2012 AV Akademikerverlag GmbH & Co. KG
Alle Rechte vorbehalten. / All rights reserved. Saarbrücken 2012

To Sabrina Norris Andrade

Preface

This book is based on my thesis. However, I have implemented some changes. These include the following:

1. I added some more recent literature to the bibliography, since the original work is from 2010.

2. The original thesis contained many notes that were actually for the correctors. These notes were removed.

3. I found some minor typing errors. Of cause, these were removed.

4. Unfortunately all figures had to be transposed to black and white. I tried to maintain the information in all of these pictures.

5. Many discussions are now up-to-date. This was specifically interesting, since nearly all of the vague assumptions from the original thesis became widely accepted theories in this field.

6. I have changed the text in such way that it is now more pedagogic.

So, I hope that the book will help the reader in understanding graphene, photoemission and / or the interaction between graphene with different substrates.

Contents

Preface ii

List of Figures vii

List of Tables xi

1 Introduction 1
   1.1 Graphene 1
      1.1.1 The band structure of graphene 4
           1.1.1.1 Derivation of the electronic structure of the graphene p_z-states 5
           1.1.1.2 Breaking the symmetry of the two carbon sublattices . 9
      1.1.2 The extraordinary properties of graphene 10
      1.1.3 Possible future applications 11
   1.2 Graphene on different substrates 11
      1.2.1 Graphene on SiC 12
           1.2.1.1 Gap or no gap 13
           1.2.1.2 Multilayer graphene that behaves like a single graphene sheet 14
      1.2.2 Graphene on metals 15
   1.3 Aims of this thesis 15

2 Experimental 17
   2.1 An introduction to photoemission spectroscopy 17
      2.1.1 The photoelectric effect 17
      2.1.2 Photoemission spectroscopy 18
           2.1.2.1 The three-step model of photoemission 18

CONTENTS

		2.1.2.2	Beyond the three-step model of photoemission	23
	2.1.3	Fermi mapping		24
	2.1.4	Core level spectroscopy		25
2.2	Apparatus			26
	2.2.1	The laboratory PES setup		26
		2.2.1.1	The UVS300 He-lamp as a photon source	28
		2.2.1.2	The functionality of hemispherical electron analyzers	28
		2.2.1.3	The goniometer	33
		2.2.1.4	Further equipment	33
	2.2.2	The BESSY endstation		34
		2.2.2.1	Synchrotron radiation	34
		2.2.2.2	Equipment of the BESSY endstation	38
	2.2.3	The ESF		39
2.3	Special issues with the experimental set-up			39
	2.3.1	Distortions in angular space		39
		2.3.1.1	Curved and straight slits	40
		2.3.1.2	Magnetic fields	45
	2.3.2	Alignment		51
	2.3.3	Ultra-high vacuum issues		54

3 Graphene on ruthenium   **57**

3.1	Introduction		57
3.2	Apparatus		58
3.3	Preparation		58
3.4	Results		61
	3.4.1	Graphene layers of different thickness on ruthenium	63
	3.4.2	Energy gap formation in graphene on ruthenium by control of the interface	66
3.5	Discussion		69
	3.5.1	The thickness of the gold layer	69
	3.5.2	The origin of the band gap	71
	3.5.3	The origin of the satellites	72
3.6	Summary		74

CONTENTS

4 Graphene on SiC produced by Nickel Diffusion — 77
- 4.1 Introduction — 77
- 4.2 Apparatus and Preparation — 79
- 4.3 Results — 81
  - 4.3.1 Core levels — 82
  - 4.3.2 Valence bands — 85
- 4.4 Discussion — 87
  - 4.4.1 The Graphene Substrate — 87
    - 4.4.1.1 Θ-Ni$_2$Si as a substrate for graphene — 89
    - 4.4.1.2 Structure and orientation of the graphene bilayer islands — 89
  - 4.4.2 Layer Thickness — 94
- 4.5 Conclusion — 95

5 Graphene on Nickel — 97
- 5.1 Graphene on nickel — 97
  - 5.1.1 A short introduction to graphene on nickel — 97
  - 5.1.2 Graphene as a spin filter — 99
- 5.2 Apparatus and Preparation — 100
- 5.3 Results and Discussion — 101
  - 5.3.1 Band maps and energy distribution curves — 101
    - 5.3.1.1 Spin filtering effects — 106
    - 5.3.1.2 Hybridization effects in detail — 107
  - 5.3.2 Fermi surfaces and constant energy maps — 110
- 5.4 Summary — 112

6 Summary, Conclusions and Outlook — 115
- 6.1 Summary — 115
  - 6.1.1 Comparison of the different growth mechanisms — 115
  - 6.1.2 Comparison of the interaction of graphene with the different substrates — 116
- 6.2 Conclusions — 117
  - 6.2.1 Possible Future Applications — 117
    - 6.2.1.1 Graphene as a spin filter — 118
- 6.3 Outlook — 118

CONTENTS

Bibliography   121

List of Figures

1.1	Some carbon allotropes.	2
1.2	Models of graphene.	4
1.3	Geometry of graphene in real and reciprocal space.	5
1.4	Theoretically calculated graphene π bands.	8
1.5	The graphene-stackings for graphene on the SiC(0001) and the SiC(000$\bar{1}$) surface.	13
2.1	The three-step model of photoemission	19
2.2	The universal curve	21
2.3	Schematic of the photoemission data acquiring method with a hemispherical analyzer and a goniometer with a β-flip.	24
2.4	Schematic demonstrating a full ARPES data set.	25
2.5	Schematic of the ARPES station in the lab.	27
2.6	A PHOIBOS150 analyzer by SPECS from two different perspectives. . .	29
2.7	Trajectories of the electrons in the electrical lens for the WAM mode with $E_p = 50\text{eV}$ and $E_k = 20\text{eV}$	30
2.8	View on the entrance plane for the same mode as in Fig. 2.7.	32
2.9	Schematic of bremsstrahlung.	35
2.10	Schematic of a synchrotron	36
2.11	Schematic of the electron trajectories in the analyzer	40
2.12	Schematic of slit-related changes in the photoemission spectra.	41
2.13	Comparison of an entrance plane with a curved slit and one with a straight slit.	43
2.14	Constant energy cuts at the Fermi surface for different θ positions for graphene on Ru(0001) taken with a straight and a curved slit.	44

LIST OF FIGURES

2.15 Magnetic field in the lens of a PHOIBOS225 analyzer by SPECS. 46

2.16 Schematic of the entrance plane with and without a field-induced shift. . 47

2.17 β-Θ scans for a curved slit and different magnetic fields in the chamber 49

2.18 Pixel-dependent K-point position for different kinetic energies, lens modes and pass energies. 50

2.19 The alignment. 51

3.1 LEED images of different Ru(0001)-based systems. 59

3.2 Overview of the photoemission intensity maps of the measured systems 61

3.3 Spectral functions and photoemission intensity profiles at the Fermi surface along Γ-K line around the K-point of graphene layers of varying thickness on Ru(0001) 64

3.4 Doping scan for K on graphene on Au on Ru(0001). 67

3.5 Comparison of the two quasi free-standing 1 monolayer graphene systems. 68

3.6 Graphene on intercalated gold on Ru(0001). 70

4.1 Different configurations of the Ni-SiC-interface after annealing. 78

4.2 LEED images and a model of graphene on Θ-Ni$_2$Si 80

4.3 Core levels 83

4.4 Valence band spectra 86

4.5 Different bilayer stackings. 91

4.6 Photoemission intensity of the bilayer band at 1.7 Å^{-1} distance from Γ-point in angular dependence. 92

5.1 Graphene on Ni(111) 98

5.2 Spectral functions of Ni(111), graphene on Ni(111) and the Ni/graphene/Ni(111) interlayer system. 102

5.3 Energy distribution curves of the samples at the high symmetry points Γ, K and M. 103

5.4 Cross sections for the nickel and carbon valence and nearby valence states. 104

5.5 EDCs at the high symmetry points for the measured systems taken with 70eV photon energy. 106

5.6 The spectral function around the K-point for graphene on Ni(111) . . . 109

LIST OF FIGURES

5.7 Fermi surfaces of Ni(111), graphene on Ni(111) and the Ni/graphene/Ni(111) interlayer system 111

5.8 Constant energy maps of one monolayer of graphene on Ni(111) 113

LIST OF FIGURES

List of Tables

4.1 Binding energies of the Ni-d-bands in different nickel silicides 88
4.2 Different possible bilayer stackings 90

GLOSSARY

1

Introduction

1.1 Graphene

Graphene has become one of the most discussed topics in physics and material science [1, 2][1]. The vast increase of publications per year is outrunning previous physics-hypes on other carbon allotropes like fullerenes and nanotubes [4]. Before the causes for this exceptional euphoria and the aims of this thesis will be described in this introduction, some elementary comments on graphene's nomenclature and its physical structure will be given.

The word *graphene* is a coinage deduced from the word *graphite* and the suffix *-ene* that is used for polycyclic aromatic hydrocarbons like naphthalene, anthracene, coronene and, in the simplest case, benzene [7]. Thus, the term *graphene* refers to one strictly two-dimensional monolayer of graphite in the (0001)-plane as shown in Fig. 1.1. The hexagonal structure that will be explained in further detail below, is not only known from graphite, but also from carbon nanotubes and, somehow, from fullerenes [8]. Therefore graphene is often used as a first approach to theoretically describe properties of these other carbon allotropes [8].

Since in graphite the distance between the graphite planes is huge (3.37Å) compared to the distance of carbon atoms within the same plane (1.42Å), graphene serves as a model particularly for the description of this three-dimensional material [9, 10]. This was first done by P.R. Wallace, who used graphene as a simple theoretical model to calculate the band structure of graphite in 1947 [9]. Wallace correctly identified graphene

---

[1]For a more recent review with a special focus on experimental physics see [3]

1. INTRODUCTION

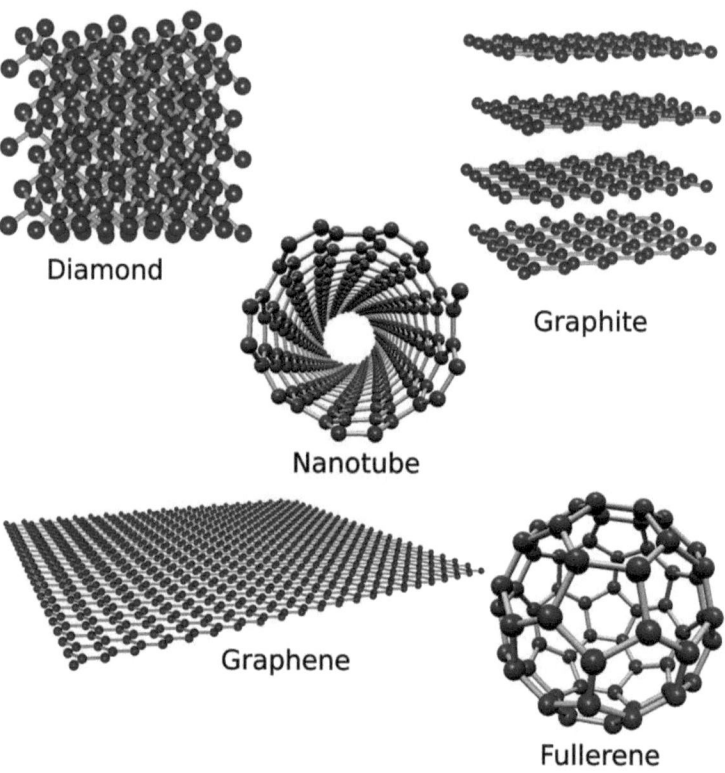

Figure 1.1: Some carbon allotropes. - Diamond and graphite have been known since prehistoric times. Nanotubes were discovered in 1991 [5]; fullerenes in 1985 [6]. The year graphene was discovered depends strongly on the point of view (more information in the text).

1.1 Graphene

as a zero-gap semiconductor and understood that the mean free path within a graphene sheet must be extraordinarily high [9]. Although more work on this system has been performed in the following decade [11, 12], only nearly 40 years later DiVincenzo and Mele pointed out that the linear dispersion of the electronic band structure near the K-point in graphene, has a zero effective mass of the charge carriers as a consequence in graphene [13]. The reason this rather simple phenomenon was not understood earlier, was the lack of interest by the scientific community. In 1937 Landau had theoretically demonstrated that strictly two-dimensional crystals were thermodynamically unstable [14], and thus scientific interest in a clearly two-dimensional crystal system such as graphene was mostly limited to theoretical modeling[1].

Still, many experimentalists have created graphene often as an unwanted byproduct in their experiments. Van Bommel et al. created graphene or thin graphite films by silicon evaporation of SiC in 1974 [15]. However, the first documented graphene synthesis as a *carbon monolayer* happened in 1977 by Oshima et al. on lanthanum hexaboride *via* segregation of dissolved carbon atoms [16]. It is particularly tricky to create clean crystal surfaces for those materials that tend to dissolve carbon atoms, and it is usually easier to study the carbon covered surfaces than to do experiments on the clean surfaces. That is probably the reason, why the second study on a clearly one-atom thick graphene layer was performed on ruthenium [17]. Subsequently there have been studies of graphene films on nickel that were grown *via* cracking of hydrocarbons [18], and on iridium using the same method [19]. However none of these studies and of those that followed in the next decade focussed on the extraordinary properties of graphene. And since most studies were actually on graphene on metal substrates graphene's properties were not preserved in any of these systems[2].

Graphene's breakthrough came with the production of free-standing graphene flakes by exfoliation from graphite crystals by Novoselov et al. [20]. On the one hand Landau's law – that strictly two-dimensional materials could not exist – was falsified, and on the other hand the extraordinary transport properties of graphene were measured in experiment for the first time [21, 22]. Similar measurements to those of Novoselov et al. had been performed in the same year for graphene on SiC by Berger et al.

---

[1] However, quasi-two-dimensional systems on substrates and 2D-electron gases as in GaAs were experimentally studied long before, but they do not contradict Landau's theorem.

[2] With SiC as the only exception. On most metals the graphene π-bands are hybridized in such way that the properties of graphene are strongly modified, as will be extensively discussed in this thesis.

1. INTRODUCTION

[23] and nearly at the same time by Zhang *et al.* [24], but the experimental investigation of graphene as the revolutionary strictly two-dimensional material, as what it is known nowadays in the science community, has often primarily been attributed to the extraordinary experimental work of Novoselov *et al.* [20][1]. In 2010 Novoselov and Geim received the noble prize in physics *for groundbreaking experiments regarding the two-dimensional material graphene.*

1.1.1 The band structure of graphene

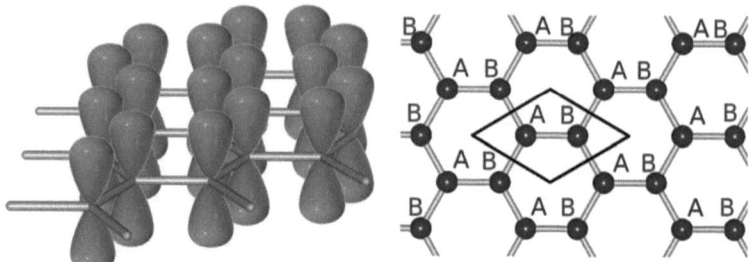

Figure 1.2: Models of graphene. - Left: The graphene p_z states are perpendicularly orientated to the graphene layer. Right: The graphene lattice with the two graphene sublattices denoted and the unit cell drawn in.

The majority of the outstanding properties of graphene are a consequence of the extraordinary band structure at the Fermi surface. Before we start the simple mathematical derivation to understand this, it is important to become aware of the fact that only the p_z electrons contribute to transport phenomena in graphene, since carbon has two electrons in the *s*-shell and four electrons in the *p*-shell; the low-lying *s*-electrons do not contribute to the transport in graphene, while three of the four *p*-shell electrons are necessary to keep the sp_2-bonds from each carbon atom to its three neighbors. Thus, the p_z orbitals that are mirror symmetric to the graphene plane, as one can see in Fig. 1.2[2], left side, are the only electrons that contribute to electronic transport phenomena.

---

[1] For further information on the controversial discussion concerning the investigation and *invention* of graphene see http://graphenetimes.com/2009/10/geim-is-not-columbus/

[2] Obviously the is a model to understand the symmetry of the states. In real graphene the p_z states are delocalized, as in benzene.

1.1 Graphene

1.1.1.1 Derivation of the electronic structure of the graphene p_z-states

The following simple derivation for the graphene π-bands is close to the one of Wallace [9], and contains some additional ideas that can be found in Saito's and Dresselhaus' book [8].

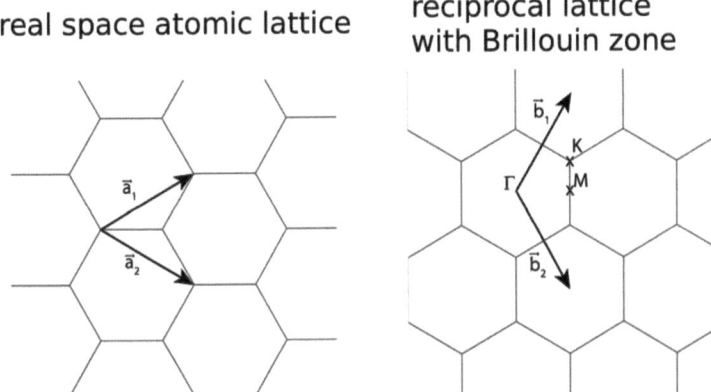

Figure 1.3: Geometry of graphene in real and reciprocal space. - The geometry of graphene with its two sublattices translates into K-space in a hexagonal lattice similar to that of a hexagonal system with one atom per unit cell.

The interatomic distance in graphene is 1.42Å which gives $a = |\vec{a}_1| = |\vec{a}_2| = 1.42\text{Å} \times \sqrt{3} = 2.46\text{Å}$ as the lattice parameter, according to Fig. 1.3. Therefore,

$$\vec{a}_1 = \left(\frac{\sqrt{3}a}{2}, \frac{a}{2}\right)$$

$$\vec{a}_2 = \left(\frac{\sqrt{3}a}{2}, -\frac{a}{2}\right)$$

which then determines the reciprocal lattice vectors as

$$\vec{b}_1 = \left(\frac{2\pi}{\sqrt{3}a}, \frac{2\pi}{a}\right)$$

$$\vec{b}_2 = \left(\frac{2\pi}{\sqrt{3}a}, -\frac{2\pi}{a}\right).$$

1. INTRODUCTION

As shown in Fig. 1.2, right side, the graphene unit cell contains two carbon atoms. These two carbon atoms are obviously inequivalent, since the three nearest neighbors of each sublattice are 60° turned with respect to each other. Therefore, in a nearest neighbor approximation with

$$\mathcal{H}_{ij} = \langle \Phi_i | \mathcal{H} | \Phi_j \rangle \tag{1.1}$$

with Φ_i and Φ_j the wave functions of states i and j, and $\mathcal{H}$ the Hamiltonian matrix element, the full Hamiltonian of our system can be written as a 2×2 matrix

$$\begin{pmatrix} \mathcal{H}_{AA} & \mathcal{H}_{AB} \\ \mathcal{H}_{BA} & \mathcal{H}_{BB} \end{pmatrix}. \tag{1.2}$$

The Bloch orbitals can then be written as

$$\Phi_j(\vec{r}) = \frac{1}{\sqrt{N}} \sum_{\vec{R}_\alpha} e^{i\vec{k}\vec{R}_\alpha} \phi_j(\vec{r} - \vec{R}_\alpha) \tag{1.3}$$

with N the number of wave vectors in the first Brillioun zone (BZ), $\alpha = A, B$ and thus $\vec{R}_A$ and $\vec{R}_B$ as the sites of the atoms of each sublattice, and $\vec{k}$ and $\vec{r}$ the wave vector and the site-coordinate in real space.

By inserting equation 1.3 into equation 1.1 for the diagonal matrix elements of the hamiltonian matrix 1.2, we get

$$\begin{aligned} \mathcal{H}_{AA}(\vec{r}) &= \frac{1}{N} \sum_{\vec{R},\vec{R}'} e^{i\vec{k}(\vec{R}-\vec{R}')} \langle \phi_A(\vec{r}-\vec{R}') | \mathcal{H} | \phi_A(\vec{r}-\vec{R}) \rangle & (1.4) \\ &= \frac{1}{N} \sum_{\vec{R}=\vec{R}'} \epsilon_{2p} + \frac{1}{N} \sum_{\vec{R}=\vec{R}'+\vec{a}_i} e^{\pm i\vec{k}\vec{a}_i} \langle \phi_A(\vec{r}-\vec{R}') | \mathcal{H} | \phi_A(\vec{r}-\vec{R}) \rangle + ... & (1.5) \\ &= \epsilon_{2p} + \text{higher terms...} & (1.6) \end{aligned}$$

with $a_i = a_1, a_2$. We assume from equation 1.5 that the main contribution to the transition matrix element comes from the nearest neighbor interaction, and therefore the energy of the $2p$ state can be approximated as ϵ_{2p}. Thus,

$$\mathcal{H}_{AA} = \mathcal{H}_{BB} = \epsilon_{2p}. \tag{1.7}$$

Therefore only the two off-diagonal matrix elements of the Hamiltonian have to be evaluated. Since every atom of sublattice A has three neighboring atoms of sublattice

1.1 Graphene

B, $\mathcal{H}_{AB}$ can be written as

$$\begin{aligned}\mathcal{H}_{AB} &= \frac{1}{N}\sum_{R,i} e^{i\vec{k}\vec{R}_i}\langle\phi_A(\vec{r}-\vec{R})|\mathcal{H}|\phi_B(\vec{r}-\vec{R}-\vec{R}_i)\rangle, i=1,2,3 \\ &= t\sum_{i=1,2,3} e^{i\vec{k}\vec{R}_i} \\ &= tf(\vec{k})\end{aligned}$$

with $t = \langle\phi_A(\vec{r}-\vec{R})|\mathcal{H}|\phi_A(\vec{r}-\vec{R}-\vec{R}_i)\rangle$. Therefore, by inserting the previously given coordinates indicated in Fig. 1.3, we get

$$f(k) = e^{ik_x a/\sqrt{3}} + 2e^{ik_x a/(2\sqrt{3})}\cos\frac{k_y a}{2}. \tag{1.8}$$

Since matrix 1.2 must be Hermitian and $f(k)$ is a complex function, we know that $\mathcal{H}_{AB} = \mathcal{H}_{BA}^*$[1]. Therefore,

$$\mathcal{H} = \begin{pmatrix} \epsilon_{2p} & tf(k) \\ tf^*(k) & \epsilon_{2p} \end{pmatrix}, \tag{1.9}$$

and the overlap integral matrix can easily be written as

$$\mathcal{S} = \begin{pmatrix} 1 & sf(k) \\ sf^*(k) & 1 \end{pmatrix} \tag{1.10}$$

with $s = \langle\phi_A(\vec{r}-\vec{R})|\phi_A(\vec{r}-\vec{R}-\vec{R}_i)\rangle$. Now, from $\det(\mathcal{H} - E\mathcal{S}) = 0$ one can find two solutions

$$E(\vec{k}) = \frac{\epsilon_{2p} \pm t\omega(\vec{k})}{1 \pm \omega(\vec{k})} \tag{1.11}$$

with

$$\omega(\vec{k}) = \sqrt{|f(\vec{k})|^2} = \sqrt{1 + 4\cos\frac{\sqrt{3}k_x a}{2}\cos\frac{k_y a}{2} + 4\cos^2\frac{k_y a}{2}}. \tag{1.12}$$

Depending on the t and s values, one can fit the theoretically calculated band structure to experimental data. Fig. 1.4 shows the theoretically calculated electronic structure of the graphene π-bands according to equation 1.11[2]. One can clearly see that the upper anti-bonding π^*-band and the lower bonding π-band touch at the K-point and render graphene a zero-gap semiconductor. This is only the case if both orbital energies ϵ_{2p} for the sublattices have the same energy. Naturally the fact that the electron wave

---

[1] The star signifies the complex conjugate.
[2] For Fig. 1.4 I used $t = 3$ and $s = 0.1$. These are reasonable values and close to the values for graphite [8].

1. INTRODUCTION

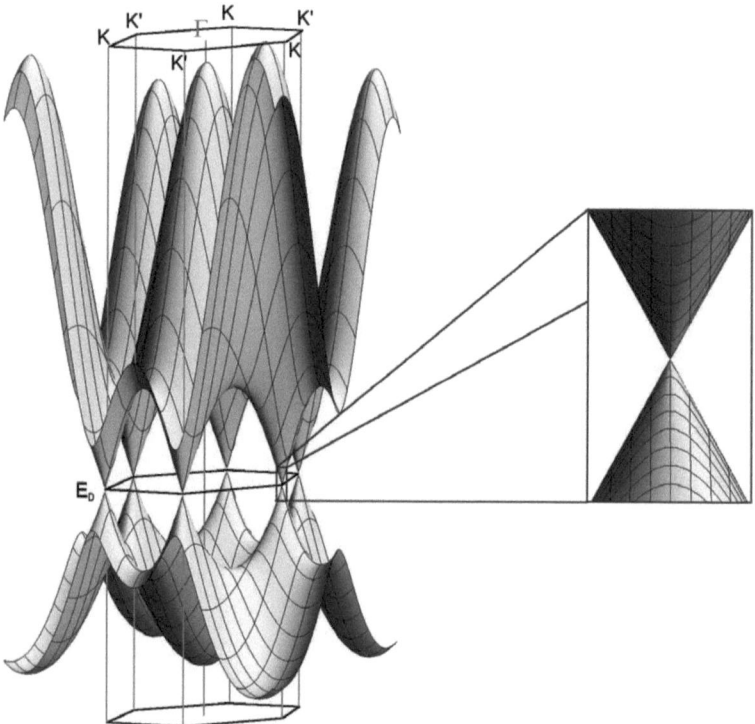

Figure 1.4: Theoretically calculated graphene π bands. - The band structure has been calculated using equation 1.11.

function's first derivation is only continuous at the K-point, when the lower π band is coupled to the upper π^* band, leads to the conclusion that the electron wave functions are now over both sublattices. Therefore, transport in graphene happens over A to B sublattice hopping and since the three nearest neighbors are at different positions for both sublattices, one has to distinguish between two different K-points K and K'.

Besides the zero-gap semi-conducting property of graphene, a remarkable feature is also the linear dispersion of the bands at the K-point near to the Fermi-level [13]. As one can see in Fig. 1.4 the two bands appear strictly conical in their structure around the K-point. Since this behavior is known for particles without any rest-mass (often referred to as Dirac-particles) these cones are often referred to as *Dirac cones* and the point where these cones meet as the *Dirac point*.

1.1.1.2 Breaking the symmetry of the two carbon sublattices

As mentioned previously the energetic equivalence of the sites of the atoms of the two sublattices is responsible for the zero-gap of the electronic band structure in graphene. A breaking of the symmetry automatically leads to a gap opening. If the energetic difference between the two sublattices is Δ, the Hamiltionian 1.9 changes into[1]

$$\mathcal{H} = \begin{pmatrix} \epsilon_{2p} + \frac{\Delta}{2} & tf(k) \\ tf^*(k) & \epsilon_{2p} - \frac{\Delta}{2} \end{pmatrix}. \tag{1.13}$$

Therefore, the band structure for the bonding and anti-bonding π states have to be rewritten as

$$E(\vec{k}, \pi^*) = \pm \sqrt{\epsilon_{2p}^2 - \frac{\Delta}{2} + t^2 |f|^2} \tag{1.14}$$

The graphene band structure around the K-point calculated with this formula results in a band gap of size Δ at the K-point and a destruction of the strictly conical band structure.

However, a gap without destroying the linear behavior of the electrons cannot be achieved in any physical system, since the wave function of any system has to be differentiable [26, 27]. Since both the anti-bonding and the bonding π-band wavefunctions, are not differentiable at the K-point and only the fact that they touch allows an electronic structure consistent with differentiable wave functions at the K-point, the Dirac-behavior of the electrons is destroyed by the appearance of a gap.

---
[1] See also Bostwick *et al.* [25]

1. INTRODUCTION

1.1.2 The extraordinary properties of graphene

The linear dispersion of the π and π^* bands at the K-point and the fact that they touch[1] leads to the conclusion that transport in graphene happens mainly by hopping of the electrons from one sublattice to the other [13]. The strictly linear band dispersion makes the electrons behave like particles without a rest mass, which travel with an *effective speed of light* $c_{\text{eff}} \approx c/300 \approx 10^6$m/s [21] through the graphene sheet. Moreover, together with the vanishing density of states this leads to an extremely high room temperature mobility of charge carriers of 15000cm^2/Vs [21], which exceeds the best values for charge carrier mobilities in silicon. The mean free path of the charge carriers in graphene has been measured to be around 0.5mm [20, 28], which is a phenomenon often referred to as *ballistic transport*. Interestingly ballistic transport in graphene is barely affected by chemical doping [29], which makes graphene a promising candidate for future field effect transistors[2].

All these extraordinary characteristics in graphene show a significantly low temperature dependence [21, 22]. The most spectacular example is probably the Quantum Hall Effect (QHE) at room temperature [30]. Moreover the QHE in graphene shows an anomalous half-integer quantization that originates from an unusual geometrical phase[3] (often referred to as Berry's phase), again induced by the strict symmetry of the two sublattices [24].

Another interesting feature of graphene as a system for fundamental research becomes clear by inserting $\epsilon_{2p} = 0$ in the Hamiltonian 1.9. Then

$$\mathcal{H} = \begin{pmatrix} 0 & tf(k) \\ tf^*(k) & 0 \end{pmatrix} \tag{1.15}$$

which is mathematically equivalent to the Pauli-matrix. Thus, the appearance of an electron in one of the two sublattices A and B is mathematically equivalent to the spin of an electron, and can therefore be referred to as a *pseudo-spin*. This makes graphene

---

[1] As mentioned in the previous subsections, the linear dispersion of course *requires* the zero band gap.

[2] Since in future field effect transistors, graphene has to be placed on a semi-conducting substrate, slight substrate-induced doping is probable. Chemical doping could correct the zero-gate voltage charge carrier density.

[3] The unusual geometrical phase in the physical case of graphene means that the wave function of an electron in graphene under a magnetic field changes its phase by $\frac{1}{2}\pi$, when describing a closed trajectory in k-space [31].

a system that allows for a study of fermionic Dirac particles [32]. Until 2009 graphene actually was the only material known to mankind that exhibits fermionic behavior for massless particles. However, topological insulators are another recently discovered system with this extraordinary property [33] In case of bilayer graphene this even leads to a system with four spin-directions [1, 2, 34].

1.1.3 Possible future applications

The high mobility of charge carriers and its low temperature- and doping-dependence make graphene a promising candidate for replacing silicon in future field effect transistor based devices [21, 34]. Indeed, industrial companies such as IBM [35, 36], Samsung [37, 38], and Intel [23, 39] are already doing research in this field, to create graphene-based field effect transistors. The biggest issue concerning such field effect transistors is the lack of an energy gap at the Dirac energy E_D (see also chapter 3).

Due to graphene's stability, it has moved into the focus of research on one-electron transistors and more generally onto the field of molecular-sized electronic devices [34, 40, 41]. Moreover, its Fermi surface that shows only electronic states at the K-points make it a promising candidate for future spin filtering devices [42, 43] (see also chapter 5) and also the first graphene-touch screen have been produced recently [44]. In summary, it seems very likely that graphene will find its way into modern electronic devices and its whole set of potential applications still has to be explored.

1.2 Graphene on different substrates

When the research for this book was done, two major issues happened to be in the focus of graphene research for industrial purposes [1]:

- The fabrication of wafer-sized graphene layers.

- The interaction of materials with graphene.

The first point relates to a very simple problem: it is clear that graphene flakes, made by the exfoliation method [20], will never find their way into industry, since those are small and rather expensive to produce [1]. So other methods of graphene production have to be explored and some will be discussed in the next subsections.

1. INTRODUCTION

The second point indicates another fundamental issue of graphene-based devices. Not only since the graphene band structure extensively depends on the symmetry of the two sublattices, one has to know how a substrate material can modify or destroy the unique properties. This is not only related to the fact that graphene sheets in devices probably will have to be placed on substrates, but also to the ineluctable electrical contacts of graphene with different materials for electronic applications.

1.2.1 Graphene on SiC

One of the best-explored methods of graphene growth on a semi-conductor is the annealing of a SiC crystal to temperatures above 1150°C, which makes the silicon atoms evaporate and leaves the carbon atoms in a graphitic structure on the surface [45, 46]. This method was first studied by van Bommel *et al.* in 1974 [15] and was rigorously researched during the second half of the 90's [45, 46, 47, 48]. The big graphene tide in research, starting in 2004, then led to studies that proved the possibility of graphene growth on SiC revealing the characteristics of free-standing graphene[1] [23, 49, 50].

However, Low-Energy Electrons Microscopy (LEEM) studies show that the quality of graphene layers grown by the previously mentioned method on the 6H-SiC(0001) surface is rather limited [51], whilst graphite films on most other surface orientations studied show less sharp Low-Energy Electrons Diffraction (LEED) spots [52]. But novel methods of graphene growth on the 6H-SiC(0001) surface seem to yield graphene samples of improved quality [53, 54], while graphene-growth on the (000$\bar{1}$) surface shows layer thicknesses that are difficult to control, but preserves the characteristics of monolayer graphene [55].

Although neither the growth of graphene on SiC(0001) *via* the classical annealing method, nor the growth of graphene on SiC(000$\bar{1}$) are special topics of this book, some information on both will be given. Before the characteristics of graphene grown on the SiC(000$\bar{1}$) surface will be discussed, the electronic structure of graphene layers grown on the (0001) surface will commented, since during the time when the experimental work for this thesis has been performed some central questions were in the focus of research.

---

[1] The term *free-standing* graphene refers to its band structure. Graphene on SiC is electron doped by 450 meV. Other possible substrate-induced changes in graphene's electronic structure will be discussed in section 1.2.1.1.

1.2 Graphene on different substrates

1.2.1.1 Gap or no gap

The first graphene layer on the 6H-SiC(0001) surface is strongly bound to the substrate and shows no linear dispersion of the graphene π bands at the K-point [56] and is thus often referred to as a buffer layer or 0th graphene layer [25]. This structure of graphene multilayer systems is also known from other substrates [57]. Thus, the second layer will be referred to as a single graphene layer in the following. This layer exhibits all important properties of a graphene monolayer with an electron-doping of 450meV [50]. However, angular-resolved photoemission spectroscopy measurements reveal that the graphene π-bands show a slightly anomalous structure around the K-point in form of a kink that is accepted as being induced by electron-phonon coupling processes [50] and a feature at the Dirac-point that has been interpreted as a consequence of electron-plasmon interactions [50] or alternatively as a substrate-induced band gap [58].

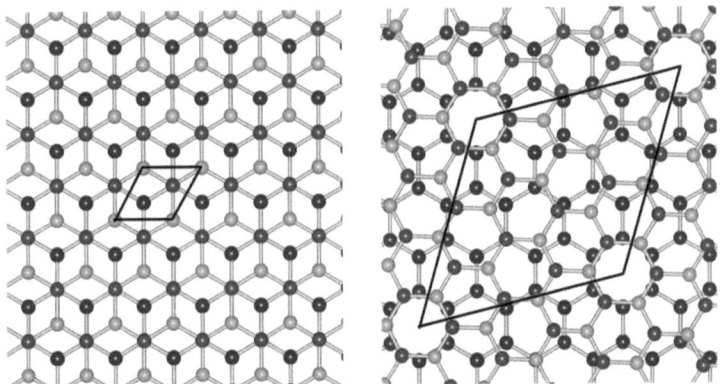

Figure 1.5: **The graphene-stackings for graphene on the SiC(0001) and the SiC(000$\bar{1}$) surface.** - A graphene bilayer in the AB-stacking (left panel) and the 27.8° rotated stacking as grown on the SiC(000$\bar{1}$) surface. Each sublattice of the upper and lower layer is shown in a different shade of grey.

From a theoretical point of view a substrate-induced band gap of a graphene layer on the buffer layer would make sense, since graphene layers are mostly stacked in the classical AB-manner as shown in Fig. 1.5. As in bulk graphite this leads to a non-vanishing Δ in the Hamiltonian 1.13 and therefore can lead to a gap opening. But if the

1. INTRODUCTION

interaction of the two graphene layers is weak enough such effect could be suppressed.

However, Zhou et al. refer to a band gap of 260meV without observing any band gap in form of a photoemission intensity minimum in their spectra[1] [58, 59]. Moreover the electron-plasmon interaction naturally results in an unusual broadening of the bands at the K-point that has been observed [50]. Zhou et al. would have to explain this kink-related broadening of the bands at E_D. This was highly controversial and has been discussed extensively [25, 60]. Such a gap would make SiC an even more promising candidate as a substrate for graphene in future electronic devices, since a band gap is still missing in such a system.

In chapter 3 of this thesis, data on a clearly gapped spectrum of a graphene layer on gold on Ru(0001) will be shown. The gap size accounts to 200meV and none of the broadening features that were seen by Zhou et al. are present[2]. As will be explained in that chapter, graphene on Ru(0001) is comparable in many aspects to the graphene on SiC(0001) system. The controversy about the gap is over nowadays and only very few people do still believe in a gap possessed by a single graphene layer on SiC.

1.2.1.2 Multilayer graphene that behaves like a single graphene sheet

Fifteen years ago the graphitization of the SiC(000$\bar{1}$) surface seemed to be a less promising approach towards waver-sized graphene sheets, since the LEED studies of Forbeaux et al. showed that the graphene planes grow in varying orientations and reveal strongly varying layer thicknesses on the same sample [61], but on the other hand these films of 10 to 20 graphene layers show the transport characteristics of single graphene sheets [28, 62]. This was attributed to the stacking of the graphene layers that differs to the usual AB-stacking, and instead are stacked in such a manner that one layer is rotated by 47.8° to the lower layer, which then preserves the two sublattices from symmetry breaking [55] (see Fig. 1.5, right panel).

Therefore this method provides graphene layers with Dirac-like quasiparticles that are easy to make. However, field effect transistors might be difficult to be based on such systems, since the charge transfer between the layers is very low and therefore the absence of field-induced change of charge carrier concentration is an issue. Moreover a gap opening could be difficult to achieve.

---

[1] Photoemission spectroscopy as a method will be explained in the next chapter.
[2] See also Enderlein et al. [57].

1.2.2 Graphene on metals

In all future graphene-based electronic applications, it is highly probable that the graphene sheets must be in contact with metals, and thus it is of essential interest to study how graphene's properties change at these points of contacts. This is particularly important since the graphene π-bands will necessarily hybridize with metallic bands on most metals[1], and in many cases this can happen near to the Fermi surface[2], as is the case for nickel [64, 65] and ruthenium [17, 66], which are both focussed upon in this thesis. In both systems the first grown graphene layer does not show Dirac-cones at the Fermi level [67, 66].

Moreover, metals are important in many methods of graphene fabrication. An often used method that will be explained in detail in chapter 5 is the cracking of hydrocarbons on the metal surface [18, 68].

1.3 Aims of this thesis

When the experimental work, as presented in this thesis, was performed, the status of graphene research made it necessary to explore different methods of graphene growth and the interaction of graphene with varying substrates. As will be explained within the next chapter, ARPES as a method particularly serves this purpose, since it makes detailed studies of the band structure of two-dimensional systems possible.

Graphene-layers of different thicknesses have been grown and studied on Ru(001) *via* segregation (chapter 3), on one monolayer of intercalated gold on Ru(001) (also chapter 3), on Θ-Ni$_2$Si(001) *via* the diffusion of nickel atoms in a SiC(001) surface (chapter 4), and on Ni(111) *via* chemical vapor deposition (CVD) (chapter 5). For each system a different method for graphene formation has been used and all three systems show significantly differing characteristics.

---

[1] Of course these hybridizations do also happen on semiconductors, as it is the case with the previously mentioned buffer layer on SiC(0001).

[2] If the hybridization happens far away from the Fermilevel, as it is the case on Ir [63], the band structure of graphene close to the Fermi surface is naturally preserved.

1. INTRODUCTION

2

Experimental

2.1 An introduction to photoemission spectroscopy

2.1.1 The photoelectric effect

The photo effect - also known as the photoelectric effect - was first observed by Alexandre Edmond Becquerel [69] and later extensively studied by Heinrich Hertz [70] and Wilhelm Hallwachs [71]. These researchers found that when a surface is exposed to electromagnetic waves above a material-specific frequency, electrons were emitted from the surface. At the end of the 19th century people tried to explain this behavior within the model of the Maxwell theory. However, this approach failed when Philipp Eduard Anton von Lenard measured in an experiment in 1902 the dependence of the energy of the emitted electrons on the frequency of the incoming electromagnetic wave [72]. According to Maxwell's theory, the energy of the emitted electrons should increase with the intensity and not with the frequency as measured be Lenard.

Albert Einstein solved this problem with a very simple model, in which the electromagnetic wave is treated as a bunch of particles, namely photons [73]. In Einstein's model, a photon excites an electron in the solid and transfers its energy completely to the electron. If the energy of the electron exceeds a certain energy which is often referred to as *work of emission* or *work function*, it may leave the surface. The kinetic energy of the emitted electron can then be written as

$$E_k = h\nu - W_f \tag{2.1}$$

2. EXPERIMENTAL

with h the Planck constant, ν the frequency of the irradiated light and W_f the material-specific work function.

2.1.2 Photoemission spectroscopy

The energy-resolved detection of photoelectrons with so-called electron energy analyzers opens the possibility of using the photo effect for spectroscopic purposes, since one would naively expect that the detected photoemission electron density would be proportional to the density of states (DOS) in the material. In fact besides the DOS many other effects influence the intensity of the photoemission signal[1].

In this case, the momentum of the emitted electron should - again naively - be able to be determined by the momentum of the electron prior to being exited by the photon, which is naturally determined by the band structure of the material[2]. Since the momentum can be measured *via* the exit angle and the energy of the emitted electron, the method is called *angle-resolved photoemission spectroscopy* (ARPES). In the following subsection a common, simplified theoretical model for the description of photoemission processes will be explained.

2.1.2.1 The three-step model of photoemission

An intuitive, but also rather incomplete model, to describe the process of photoemission is given by the semi-classical three-step model [74, 75]. In contrast to offering a complete description, the photoemission process within this model is strictly divided into three consecutive steps, as demonstrated in Fig. 2.1 a).

- The electron is excited by the photon.
- The electron moves through the material to the surface.
- The electron escapes the potential barrier of the surface.

The momentum of the photon is usually neglected within this model. This is reasonable for all photoemission studies that are presented in this thesis. Therefore, the

---

[1] i.e. the transition probability of the initial state to the final state of the electrons. See also subsection 2.1.2.2.

[2] Here one would have to take into account possible quasi-particle processes within the crystal, as well as interference effects of the emitted electron wave. See also subsection 2.1.2.2

2.1 An introduction to photoemission spectroscopy

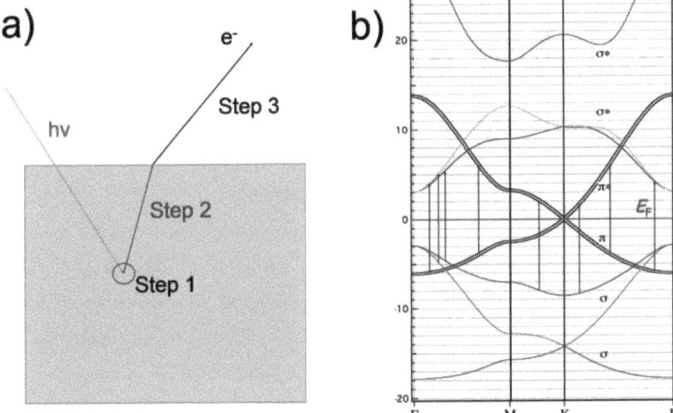

Figure 2.1: The three-step model of photoemission - a) a schematic to demonstrate the three steps of photoemission in this model. b) Band structure of graphene taken from Saito and Dresselhaus [8], as presented in a lecture of Eli Rotenberg. The arrows mark allowed transitions of electrons from occupied to non-occupied states, as required for step 1.

2. EXPERIMENTAL

momentum $\hbar\vec{k}$ of the electron must be conserved. Thus,

$$\vec{k}_f = \vec{k}_i + \vec{G} \tag{2.2}$$

with $\vec{k}_f$ and $\vec{k}_i$ as the wave vectors of the electron in the final and the initial state, while $\vec{G}$ is the reciprocal lattice vector. Like in an atom transitions with $\vec{k}_f = \vec{k}_i$ are forbidden, but the periodical potential allows quasi-perpendicular transitions, since in the one-Brillouin Zone picture the electron does not gain momentum. Since energy conservation must be fulfilled, we can write for the energy in the final state

$$E(\vec{k}_f) = E(\vec{k}_i) + h\nu. \tag{2.3}$$

This means the electron gains the energy $h\nu$ from the photon without gaining momentum. Thus, the electron becomes excited into an upper band without changing its momentum as demonstrated in Fig. 2.1, right panel.

The next step is the transport of the electron through the material. Since this can be considered as a classical transport process, scattering events will inevitably occur. By definition, scattering events will change the momentum and energy of the electron and thus destroy the information carried by them. Therefore, non-scattered electrons will give a sharp photoemission signal from the band structure of the material, while scattered electrons will either obtain a broadening of the photoemission signal or appear as background signal in the data sets. Thus, the sharp photoemission signal will come from the electrons that originated close to the surface.

The average distance an electron can travel within a material without scattering events is called the *mean free path*. The mean free path depends strongly on the electron energy, but is relatively independent of the material the electron goes through as one can see in the so-called universal curve in Fig. 2.2. Since the electrons that were detected during the experiments in this thesis usually had kinetic energies ranging from 50 to 100eV, only electrons of the upper few layers contributed to our measurements. This makes ARPES a highly surface sensitive method and thus the perfect tool to examine two-dimensional structures such as one- or few-layer graphene.

The last step of the three-step model is the escape of the electron from the surface. If the surface can be considered as flat, $\vec{k}_f$ can be split in two components: $\vec{k}_{f\parallel}$ parallel

2.1 An introduction to photoemission spectroscopy

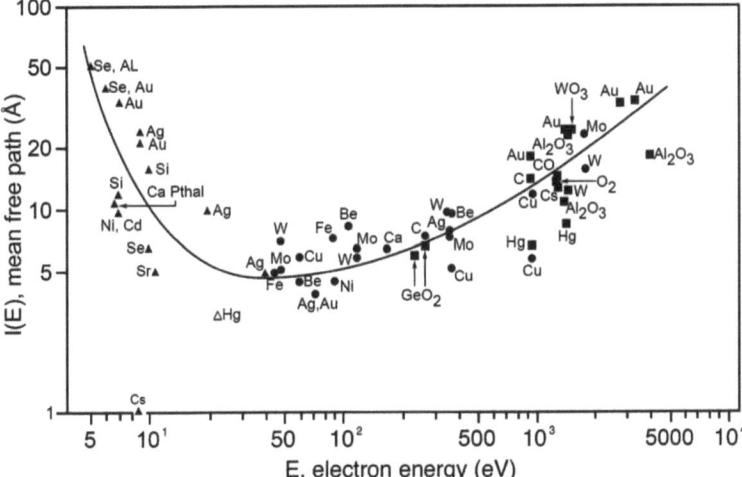

Figure 2.2: The universal curve - The markers represent the experimentally determined mean free paths. After Zangwill's *Physics at surfaces* with experimental data from Rhodin and Gadzuk [76], and Somorjai [77]. The theoretical curve is from Penn [78].

2. EXPERIMENTAL

to the surface and $\vec{k}_{f\perp}$ perpendicular to the surface. If $\vec{q}$ is the wave vector of the electron outside the crystal, then

$$\vec{q}_{\parallel} = \vec{k}_{f\parallel} + \vec{g}_{\parallel} \qquad (2.4)$$

with $g_{\parallel}$ as the reconstructed reciprocal lattice vector of the surface plane. In case of (100), (010) and (001) surfaces, this will be the reciprocal lattice vector $\vec{G}$ of the specific crystal lattice in the specific direction[1].

In contrast, $\vec{k}_{\perp}$ is not conserved during the process, since the electron must lose momentum when leaving the surface. The experimental determination of the spectral function in the $\vec{k}_{\perp}$ direction is far more complicated to measure than $\vec{k}_{\parallel}$. It has to be performed either by changing the photon energy or by measuring different orientations of the same crystal. For both approaches the resolution is rather limited, however since graphene is a strictly two-dimensional material no spectral function measurements in the $\vec{k}_{\perp}$-direction had to be performed for this thesis.

The kinetic energy of the emitted electron is determined by

$$E(\vec{q}) = \frac{\hbar^2 \vec{q}^2}{2m} \qquad (2.5)$$

with m the rest mass of the electron. Now, if Θ is the exit angle of the emitted electron, $\vec{q}$ can be written as $\vec{q} = \frac{\sin\Theta}{|\vec{q}_{\parallel}|}$ and thus equation 2.5 transforms to

$$\sqrt{E(\vec{q})} = \frac{\hbar |\vec{q}_{\parallel}|}{\sqrt{2m} \sin \Theta}. \qquad (2.6)$$

By multiplying with $\vec{q}_{\parallel}$ and inserting equations 2.2 and 2.4[2], we get

$$\vec{k}_{\parallel i} = \sqrt{\frac{2mE(\vec{q})}{\hbar^2}} \sin \Theta \frac{\vec{q}}{|\vec{q}_{\parallel}|} - \vec{G}_{\parallel} - \vec{g}_{\parallel}. \qquad (2.7)$$

Since all measurements presented here were performed on graphene or unreconstructed surfaces, thus allowing the assumption of $\vec{G}_{\parallel} = \vec{g}_{\parallel}$, the formula can be further simplified for transforming angular-resolved data sets into k-space.

$$k_{\parallel i} = \sqrt{\frac{2mE(\vec{q})}{\hbar^2}} \sin \Theta, \qquad (2.8)$$

---

[1] In case of reconstructed surfaces such as (531) $g_{\parallel}$ will be the reconstructed lattice vector of the surface.

[2] This is possible since the momentum in parallel direction is preserved in our model.

2.1 An introduction to photoemission spectroscopy

which is the formula that has been used for all transformations from *angular* space into k-space that were done in this thesis.

Since the kinetic energy is a photon energy-dependent variable and band structures do not intrinsically depend on this value, the measured spectra are usually mapped against the binding energy $E_b = E_F - E(\vec{k}_i)$. Since logically $E(\vec{q}) = E(\vec{k}_f) - E_V$, with E_V as the vacuum potential, together with equation 2.3 we get

$$E_b = h\nu - \Phi - E(\vec{q}), \qquad (2.9)$$

where $\Phi = E_V - E_f$ is the *work function*. As a result, within this simplified model, a two-dimensional band structure can be completely mapped by only measuring the exit angle and the energy of the emitted photoelectrons.

2.1.2.2 Beyond the three-step model of photoemission

The previously described model neglects not only the momentum of the photon but also all kinds of quasi-particle interactions within the crystal, although these interactions can significantly influence the measurement [50, 79, 80]. A complete description of a real photoemission process is given by *Fermi's golden rule* that generally gives a way to calculate the transition rate from one quantum state to a continuum of states. For a photoemission process the transition rate can then be written as

$$j(\vec{R}, E, h\nu) \propto \sum_{i_{oc}} |\langle \Phi_f | \delta H | \Phi_i \rangle|^2 \delta(E_i - h\nu - E_f) \qquad (2.10)$$

with Φ_f and Φ_i as the wave functions of the initial and the final state and δH as the dipole operator. The sum is over all occupied states. This formula completely describes the photoemission process and is usually difficult to solve. However, in most cases the transformation of pure ARPES data within the three-step model is sufficient to create band maps that are correct within the accuracy of measurement. When multi-particle interactions like electron-plasmon or electron-phonen coupling influence the measured data in high-resolution ARPES-studies and the transformation into the band maps is done within the three-step model, the additional effects appear as deformations in the band maps [81, 82].

2. EXPERIMENTAL

Since in all data presented in this work multi-particle processes, such as electron-phonon or electron-plasmon interactions play only a minor role[1], a further discussion is out of the scope of this thesis.

2.1.3 Fermi mapping

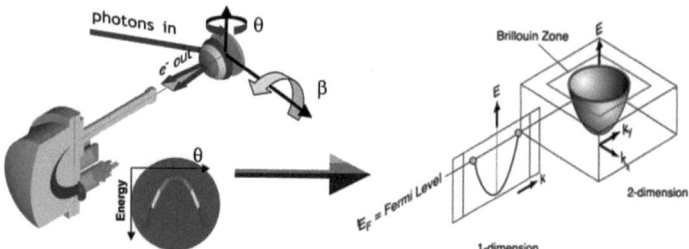

Figure 2.3: Schematic of the photoemission data acquiring method with a hemispherical analyzer and a goniometer with a β-flip. - The analyzer is able to resolve the incoming electrons in terms of energy and θ-angle. Using equation 2.8 the resulting spectrum can be easily transferred to an energy vs. one k-direction spectrum. By flipping the goniometer along the β-angle, full k-space mapping is possible by stacking together the data sets for different β-positions. The schematic has been taken from Eli Rotenberg.

Hemispherical electron analyzers of the most recent generation are able to resolve the energy and one component of the exit angle of the emitted photo electrons. This is the case for all analyzers used for this thesis, namely the PHOIBOS100 and PHOIBOS150 by SPECS^{TM} as well as the R4000 by Scienta^{TM}. As a convention we will refer to the angle which is covered just by the analyzer as θ, as is also demonstrated in Fig. 2.3. If the sample is mounted on a goniometer that can be turned around the θ-axis, full two-dimensional band mapping along one line in k-space is possible. However, to map the *full* two-dimensional band structure and to measure the k-dependent DOS at the Fermi level (this is called Fermi mapping), meaning *all* points in k-space for arbitrary photon energies, another axis has to be added.

---
[1]With one exception concerning a minor issue in the evaluation of the data presented in Chapter 3.

2.1 An introduction to photoemission spectroscopy

In the past, such photoemission data sets have always been acquired by turning the sample around its azimuth ϕ and take bandmaps along the θ-direction [83, 84]. Due to their circular appearance these spectra are colloquial referred to as *pizza plots*. These plots have several disadvantages: the k-space resolution decreases with ascending distance to the Γ-point, a sample mounted in a non-perpendicular manner will make it impossible to measure EDCs at the Γ-point, and it is very tricky to measure inhomogeneously covered samples. The first point in particular makes this scanning method very inefficient for graphene band mapping, since the region of interest for this material is usually at the K-point.

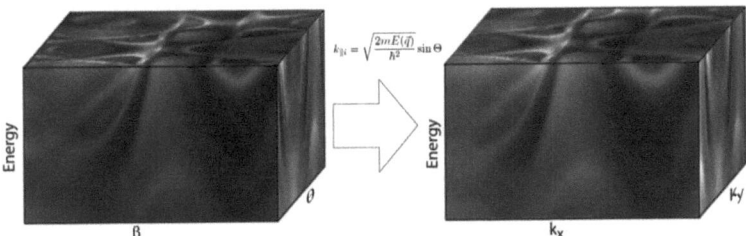

Figure 2.4: Schematic demonstrating a full ARPES data set. - The data sets are represented in a 4-dimensional matrix.

A much better method of Fermi mapping is performed by adding a *flip* axis β as demonstrated in Fig. 2.3. This axis allows acquiring two-dimensional band maps without any disadvantages. In Fig. 2.4 a full data set in angular space and in the k-space presentation is demonstrated. By including both $\vec{k}_\parallel$ directions, the energy and the photoemission intensity data sets are saved in 4-dimensional matrices.

Further details concerning the photoemission stations used will be given in section 2.2.

2.1.4 Core level spectroscopy

Our valence band spectra are mostly taken with photon energies below 100eV, since low photon energies give a better energy- and $k_\parallel$-resolution. However, when higher

2. EXPERIMENTAL

photon energies[1] are used, one can excite the electrons from the lower lying energy levels. These electrons usually do not contribute to the bonding of the crystal atoms and show no or only weak dispersion, since they are in the closed shells of the element. Commonly they are referred to as core electrons, and thus their detection is called *core level* spectroscopy.

Core levels change their energy depending on the chemical surrounding. Therefore core level spectroscopy is a very useful tool to determine the different chemical configurations of one element that has differing sites in a crystal, and to detect modifications of the crystal structure.

2.2 Apparatus

Three different ARPES-stations have been used for acquisition of data presented here: One synchrotron-independent ARPES station with a He-lamp as a photon source, one ARPES station for exclusive usage at BESSY and furthermore, for the data presented in chapter 3, another ARPES station at the ALS. In the following subsections all three setups and their components will be described in detail. This section might be particularly useful for people who are new to Fermi-mapping or want to set up an ARPES-station for Fermi-mapping by themselves.

Since electrons interact with gases, all stations were always kept under ultra-high vacuum conditions. This was attained with a combination of membrane pumps, turbo molecular pumps (TMPs), ion pumps and titanium sublimation pumps (TSPs). Further information concerning ultra-high vacuum techniques will be given in subsection 2.3.3.

2.2.1 The laboratory PES setup

As a constantly usable ARPES station, we set up an apparatus with a UVS300 He-lamp by SPECS^{TM} as a photon source and a PHOIBOS150 electron analyzer. A schematic of the chamber is shown in Fig. 2.5. The goniometer is moveable in all three spatial directions by several centimeters. Moreover the mechanics allow the sample to be rotated in θ- and β-direction as indicated in Fig. 2.5 in the two right subfigures. The ϕ axis can only be moved in a certain fixed β-position.

---

[1] For core-level measurements that have been acquired for this thesis photon energies between 150 and 1000eV were used.

2.2 Apparatus

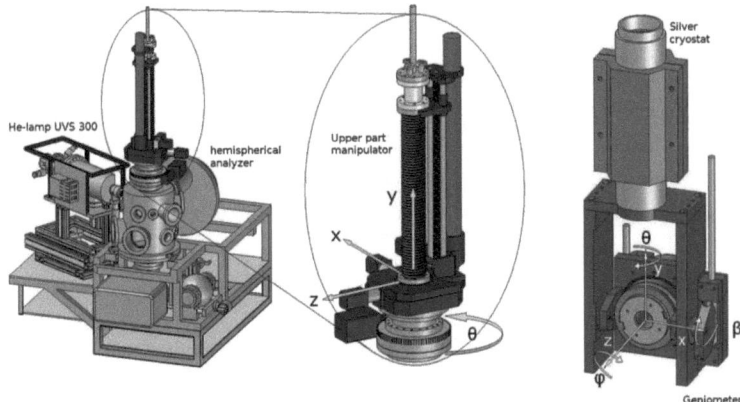

Figure 2.5: Schematic of the ARPES station in the lab. - The UVS300 serves as a photon source. The goniometer is mounted on the lower end of the cryostat. A differentially pumped rotary feedthrough allows rotation of the entire manipulator apparatus around the θ-axis. The goniometer is mounted on the lower part of the manipulator and allows β-flip movement, as well as azimuthal rotation (angle ϕ). The axes x and z rotate with θ. The picture was made by B. Frietsch.

2. EXPERIMENTAL

2.2.1.1 The UVS300 He-lamp as a photon source

To create high-intensity ultra violet (UV) radiation we used the UVS300 He-lamp by SPECS^{TM}. For operating conditions, a helium atmosphere of 10^{-5}mbar was stabilized within the lamp. The electrons, emitted by a titanium filament, are guided *via* a strongly inhomogeneous magnetic field to the discharge section, where a He-plasma is generated. In this ARPES station we use a TMM302 grating, also by SPECS^{TM}, to separate the resulting He-typical radiation of the He I and He II line (21.22 and 40.82eV). To create a focused spot with a diameter of 1mm on the sample, a metal capillary is used.

Although synchrotrons[1] have the advantage of tunable photon energy, huge photon energy range, extremely high photon flux and a small beam focus, a He-lamp is still superior in some details. Aside from the fact that an ARPES-apparatus in a lab is always usable[2], the narrow line width of the He-emission lines gives a better energy resolution than most state-of-the-art synchrotron monochromators can provide.

However, a He-lamp limits measurements to two photon energies, which makes these sources useless for a wide range of photoemission experiments[3].

2.2.1.2 The functionality of hemispherical electron analyzers

Not only the PHOIBOS150, but all other 180° hemispherical electron analyzers that were used in the scope of this thesis, essentially function by the same basic mechanism. Some basic principles will be explained in this subsection. For further information the book of Granneman and van der Wiel is recommended [85]. However, most of the principles explained here are from the manual for the PHOIBOS100 and PHOIBOS150 analyzer by SPECS^{TM}.

A hemispherical analyzer consists of an electrical lens for the angular resolution and a hemisphere to resolve the energy. In Fig. 2.6 a PHOIBOS150 analyzer is shown. The photoelectrons enter the analyzer through the nose cone into the lens. To cut off scattered electrons, SPECS analyzers are equipped with an iris. After the electrons were angularly or spatially resolved in the lens, they enter the hemisphere, which then

---
[1]The functionality of a synchrotron will be explained in subsection 2.2.2.1.

[2]Beam time at synchrotrons is very limited and periods of beam time are mostly limited to one to four weeks with very demanding beam time schedules.

[3]i.e. resonant photoemission spectroscopy.

2.2 Apparatus

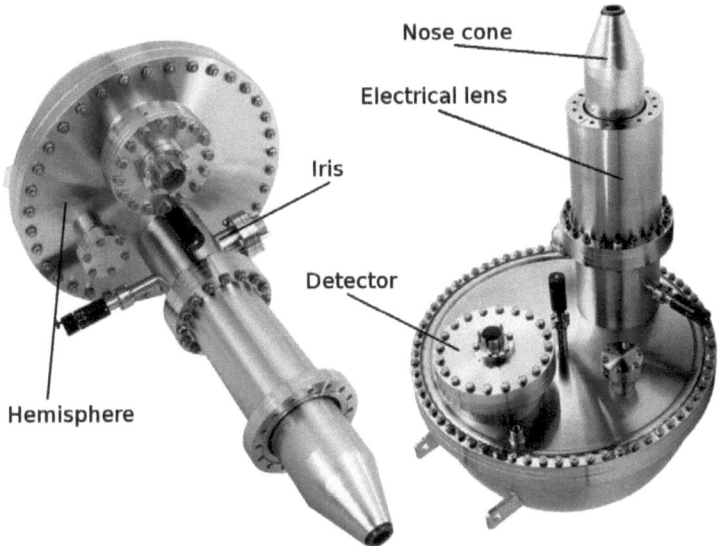

Figure 2.6: A PHOIBOS150 analyzer by SPECS from two different perspectives. - Some of the most important parts are labeled. The image was taken from SPECS™.

2. EXPERIMENTAL

resolves their energy. At the end of the electrons' trajectories they are detected by a channeltron or channel plates. The latter have been used for all spectra presented in this thesis.

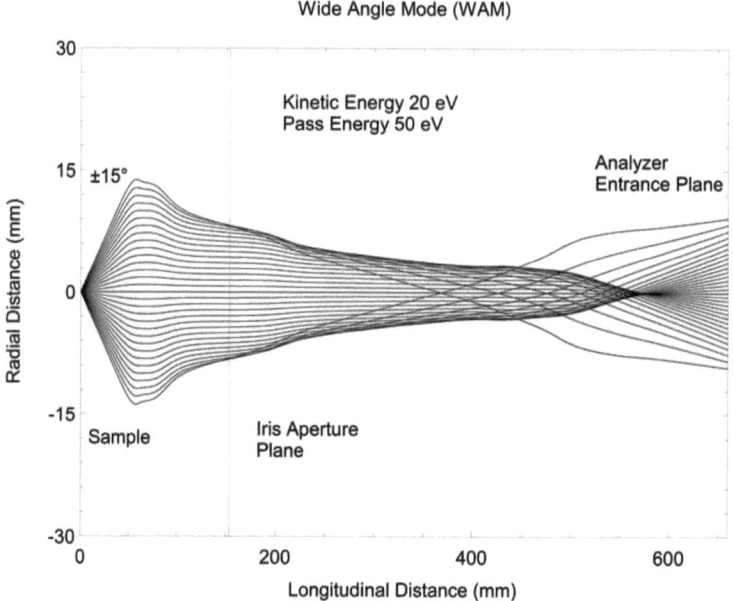

Figure 2.7: **Trajectories of the electrons in the electrical lens for the WAM mode with $E_p = 50\text{eV}$ and $E_k = 20\text{eV}$** - According to equation 2.11 the retarding ratio is 2/5. The image was made by Sven Mähl.

The strictly cylindrically-symmetric electrical lens consists of an array of electrical potentials that serve to widen and refocus the electron beam, which then hits the analyzer entrance plane as demonstrated in Fig. 2.7. Although there exist lens modes that allow spatial resolution, we will only focus on the angle-resolving lens modes in this subsection. These lens modes let the lens bundle the electron beam in a manner that makes the electrons hit on the entrance slit plane in such way that the emission angle from the sample and the distance from the beam center on the entrance slit plane obey an isomorphic relation to each other. In the best case this relationship is linear and the proportionality factor is given by the magnification. However, the so-called wide-angle mode (WAM) that allows the taking of scans within an angular range of 26°, cannot necessarily provide a linear relationship over the whole angular window. This is particularly the case if the retarding ratio (explained in detail later in this section) is close to non-defined regions.

The electrons that leave the material with a certain kinetic energy E_k are retarded within the lens down to the *pass energy* E_p. Electrons with the same E_k to E_p ratio will have equal trajectories and thus

$$R = \frac{E_k}{E_p} \tag{2.11}$$

is called the retarding ratio. Lens modes have to be designed for different retarding ratios and therefore every lens mode can only be used within a certain interval of retarding ratios. For most scans that have been performed for data acquisition presented herein, retarding ratios between 0.5 and 2 have been used.

In the exit slit plane, the exit slit cuts off a certain interval of the β-direction, as demonstrated in Fig. 2.8. Since the spectra taken in angle-resolving lens modes are usually barely effected by huge spot sizes, the entrance slit size primarily determines the resolution in β-direction.

Behind the entrance slit, the electron beam enters the hemisphere in which an electrical field guides the electrons. Only electrons that have a kinetic energy within a certain interval will reach the exit plane. Electrons with higher energies will be absorbed in the outer hemisphere and electrons with lower energies will be absorbed in the inner hemisphere. An electron with the nominal pass energy will pass the hemisphere in the centered orbit. Thus,

$$E_p = ek\Delta V \tag{2.12}$$

with e the electron charge, k a calibration constant which depends on the radii of the outer and inner hemisphere, and ΔV the potential difference between inner and outer hemisphere. This means that E_p is proportional to ΔV which determines also the kinetic energy interval of electrons that can pass through the hemisphere[1].

---

[1] Of course this is only valid, if the slit size is neglected. However, since for most scans shown in this thesis a narrow slit (10 to 20μm width) has been used, this effect can be neglected here.

2. EXPERIMENTAL

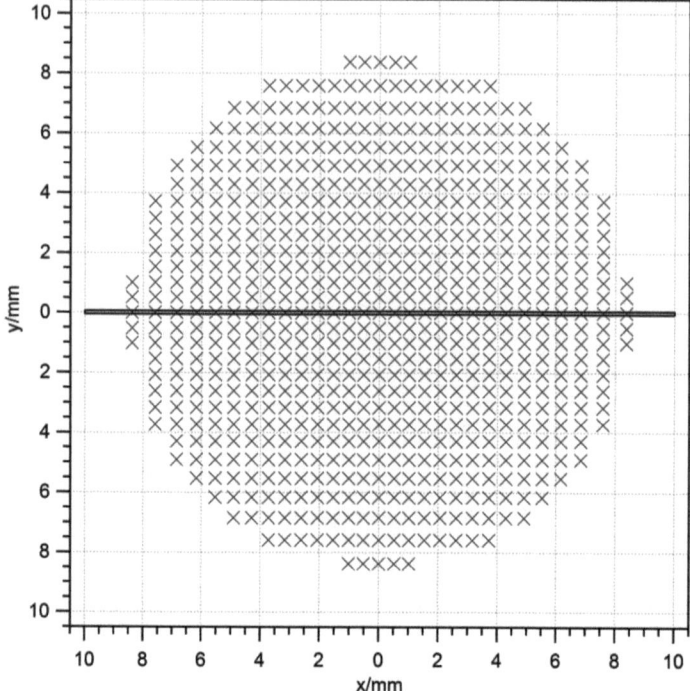

Figure 2.8: View on the entrance plane for the same mode as in Fig. 2.7. - x and y represent the two orthogonal spatial directions in the entrance plane. The y-coordinate can be directly transformed into the β coordinate with an isomorphic function, as the x-coordinate can be transformed into the θ coordinate. Every cross symbolizes an electron that left the sample in the angular direction (Θ, β) with Θ and β running from 0 to 15° in 1° steps. The black rectangle symbolizes the slit. 0.1×20mm was a typical slit size used for the experiments done for this thesis.

2.2 Apparatus

On the exit plane of the hemisphere the electrons are detected by so-called microchannel plates. These channel plates consist of an array of electron multipliers each with a diameter in the micrometer range.

The PHOIBOS150 hemispherical electron analyzer's center trajectory has a radius of 15cm. It has an identical lens as the PHOIBOS100, which can be run in several angular-dispersive modes namely the HAD (High-Angular Dispersion) mode with a nominal angular acceptance of 6°, the MAD (Medium-Angular Dispersion) mode with an angular acceptance of 8°, the LAD (Low-Angular Dispersion) mode with an angular acceptance of 14°, and the WAM (Wide Angle Mode) with an angular acceptance of 26°.

It should be mentioned that the denoted angular acceptance, as given by the manufacturer, describes the minimum angular acceptance guaranteed for the specific lens mode. Several factors like the retarding ratio and the exact distance from the sample to the analyzer influence the width of the angular window actually available. This can best be seen in Fig. 2.8; the crosses stand for electrons which come from the sample from -14 to $+14°$ exit angle, which already exceeds the denoted scope of the WAM.

2.2.1.3 The goniometer

To make full Fermi mapping possible a goniometer with β-flip, designed by B. Frietsch and K. Horn, is used. As demonstrated in Fig. 2.5, the goniometer has six axes. The samples are mounted on molybdenum sample holders that have two separated contact plates. One of the contact plates is insulated against the rest of the goniometer, which makes direct current heating possible. The sample can be cooled with a flow cryostat. Copper braids running from the cryostat foot to the back of the goniometer guarantee good heat transfer. A silicon diode is mounted behind the sample to control the temperature. Temperatures down to 15K were reached, when using helium as coolant.

2.2.1.4 Further equipment

The lab chamber is further equipped with a low-energy electron diffraction (LEED) spectrometer to control sample preparation. LEED is a common supporting technique for sample preparation in modern surface science experiments. Electrons of low energy

2. EXPERIMENTAL

are diffracted from the surface and hit a fluorescent screen on which the reciprocal lattice with its lattice points is then projected [86].

Additionally in the lab chamber a transfer arm is mounted behind a gate valve to exchange samples without venting. Several other gate valves are mounted on the chamber to make the installation and exchange of evaporators possible without venting.

The chamber is rough-pumped by a simple membrane pump which creates a pre-vacuum for the turbo molecular pump (TMP). For further vacuum improvement, a titanium sublimation pump (TSP) and an ion getter pump are additionally operational.

2.2.2 The BESSY endstation

The chamber that was used for the major part of this work is set up on a long-term basis at the Berlin storage ring for synchrotron radiation (BESSY). In contrast to most synchrotrons, BESSY has only a few end stations that are left set up continuously on their beamlines. Usually synchrotron radiation users bring their own end stations and mount them on a beam line, which can then be used as a photon source. Thus, although our endstation remained at the synchrotron, it was mounted on several different beam lines.

2.2.2.1 Synchrotron radiation

If a charged particle is accelerated by an external electric field, radiation (bremsstrahlung) is emitted as demonstrated in Fig. 2.9. For the specific case of bremsstrahlung that gets emitted in an x-ray gun, when an electron beam hits an anode, a classical description was given by Kramers in 1923 [87]. However, actually the effect results from Maxwell theory and is not required by classical energy and momentum conservation. This becomes particularly clear if the electric field is exchanged by a magnetic field which classically does not influence the energy of a particle. However, a charged relativistic particle which is under the influence of a magnetic field still emits bremsstrahlung. This specific type of bremsstrahlung is called synchrotron radiation.

Thus, within relativistic theory a charged particle with a spiral trajectory in a magnetic field constantly loses energy that is then transferred to the photons emitted.

2.2 Apparatus

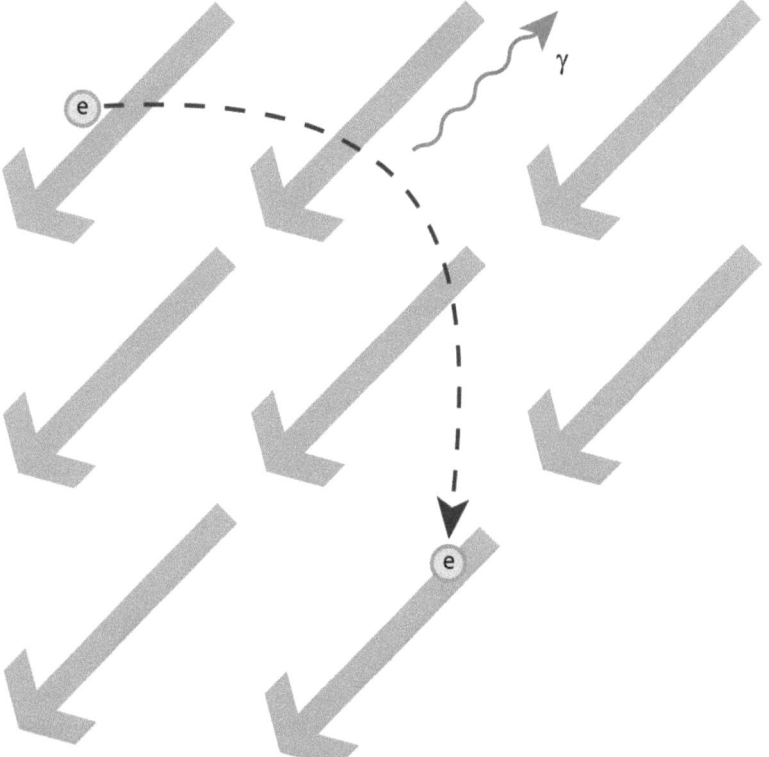

Figure 2.9: Schematic of bremsstrahlung. - A charged particle gets accelerated by an external field and emits a photon.

2. EXPERIMENTAL

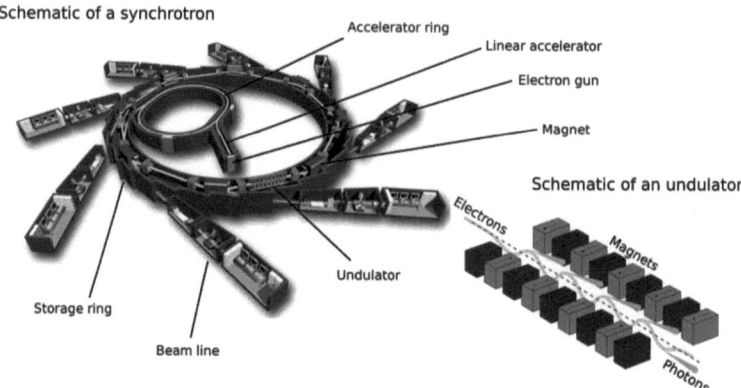

Figure 2.10: Schematic of a synchrotron - The schematic shows all basic parts of a synchrotron and an undulator. The models were taken from wikipedia (http://de.wikipedia.org/wiki/Synchrotron and http://de.wikipedia.org/wiki/Undulator).

During a full circle with the radius R an electron loses the energy

$$\Delta E = \frac{e^2 \beta^3 \gamma^4}{\epsilon_0 3 R}, \qquad (2.13)$$

with $\beta = v/c$, where v is the velocity of the electron and c the speed of light in vacuum, while γ represents the Lorentz factor $1/\sqrt{1-\beta^2}$. As a result, more photons will be emitted by the electrons if their velocity is closer to the speed of light. Furthermore, this means that if the electrons are supposed to stay on the same trajectory in every subsequent circle, they have to be regularly accelerated by electric fields. If the electrons are accelerated by an electric field in a constant magnetic field and therefore describe a spiral trajectory, the accelerator is called *cyclotron*. If the electrons are accelerated by an electric field and guided by a magnetic field and these two fields are synchronized in such way that the electrons describe a trajectory with a constant radius, the accelerator is called *synchrotron*. In BESSY and most other state-of-the-art devices for synchrotron radiation the accelerated electrons are guided into a *storage ring*, in which the electrons are kept at constant energies and therefore naturally describe a trajectory with a constant radius.

As demonstrated in Fig. 2.10, before the electrons are guided into the accelerator ring they are emitted from an electron gun, usually consisting of a cathode and an anode, and are then linearly accelerated. To reach the end velocity the electrons are then further accelerated in the synchrotron and then guided into the storage ring.

In the BESSY storage ring the particles do obviously not describe a really circular trajectory but travel usually straight until they reach a magnet as a deflecting device. Then within a very small spatial interval synchrotron radiation is emitted. In some more complex deflecting devices (usually with a tunable deflection), the photon beam is then guided through optical elements to a monochromator that leads the subsequent monochromatic beam into the beam line.

All measurements presented in this thesis that used a device for synchrotron radiation as a photon source were performed on beam lines with an undulator as a deflecting device. The schematic is shown in Fig. 2.10; the electron beam gets deflected several times by a chain of alternately polarized dipole magnets. The distance from one dipole magnet to the next with the same polarization is called λ_U. This arrangement of a deflecting device leads to a strong beam intensity and relative good energy resolution. The parameter that determines these characteristics is

$$K = \frac{eB\lambda_U}{2\pi m_e c}, \qquad (2.14)$$

where e is the electron's charge, B the magnetic field, m_e the rest mass of en electron and c the speed of light. K-values lower then one provide narrow energy bands, since the amplitude of the electron trajectory is low. High K-values lead to higher beam intensity with poor energy resolution. If the undulator has $K \gg 1$, it is referred to as a *wiggler*. The high amplitudes of the electron trajectories often make it necessary to use a Halbach-array[1] arrangement for the dipole magnets instead of the simple alternating arrangement as shown in Fig. 2.10.

In an undulator and in a wiggler one can tune the synchrotron radiation in terms of energy and polarization[2] by varying the magnetic field as well as the orientation and position of the magnets.

A synchrotron radiation device as a photon source has several advantages in comparison with a He-lamp or an x-ray gun:

---
[1] A Halbach-array refers to an array with the structure ↓←↑→↓← ...
[2] Synchrotron radiation is linearly polarized in the center of the radiation cones. Further from the center it is circularly polarized.

2. EXPERIMENTAL

- A wide continuous range of different photon energies can be used[1].

- Intensity and luminosity are higher than from other photon sources, except LASERs.

- Well defined and tunable linearly and circularly polarized light.

- The radiation is pulsed, which opens a wide range of possibilities in measuring time-dependent processes.

BESSY ll is a high-end synchrotron in which the electrons are accelerated to 1.7GeV. When the measurements presented here were made, BESSY had in its ordinary operation mode (namely the multi bunch mode) 350 electron packets in the ring with a time-wise distance of 2ns. The ring current is then between 250 and 300mA. Furthermore, BESSY sometimes runs in the single bunch mode with one single electron packet, making time-resolved measurements possible[2]. Additionally, the low-alpha mode with compressed electron packets allows measurements with terahertz radiation.

2.2.2.2 Equipment of the BESSY endstation

The BESSY endstation basically consists of two vacuum chambers.

The analysis chamber that is attached to the beam line has a goniometer which is identical in construction to the one in the lab apparatus. It uses a PHOIBOS100 analyzer that is different from the PHIOBOS150 by the 5 cm smaller diameter of the mean electron trajectory inside the hemisphere. In addition, the analysis chamber contains a LEED, several gate valve protected ports that allow the installation of evaporators, and a separated heating stage. By using an especially shaped wobble stick, the sample holder can be placed on the heating stage. Afterwards a small filament can be moved right behind the sample and by applying high voltage (HV) to the heating stage and switching on the filament, electron-beam heating can be performed.

The analysis chamber is pre-pumped by a membrane pump and a small TMP. Another TMP, an ion pump and a TSP are operational in the chamber.

The preparation chamber is equipped with a simple 4-axes goniometer (with axes x, y, z and θ), a LEED, a sample garage for up to three samples and a load lock. The possibility of preparing a sample and simultaneously doing photoemission scans

---

[1] i.e. from 50 to 1500eV
[2] During single bunch mode the electron packages have a time-wise distance of 800ns.

on a different sample in the main chamber makes this end station very suitable for synchrotron beam times which are usually time-wise limited.

The preparation chamber is rough-pumped by a membrane pump and has both a TMP and a TSP.

2.2.3 The ESF

The measurements presented in chapter 3 were all done at the electronic structure factory (ESF) on beam line 7 at the Advanced Light Source (ALS) in Berkeley, California. During the stay of the author in Berkeley, the ESF consisted of an analysis chamber, a preparation chamber, a storage chamber with space for over 50 sample holders, and two small chambers with sample transport and storage systems.

Aside from several evaporators the analysis chamber has a full 6-axes goniometer which can be cooled down to temperatures below 17K, an R4000 hemispherical analyzer by Scienta, and a LEED spectrometer. Additionally a He-lamp is mounted, which makes ARPES measurements without synchrotron radiation possible.

Eli Rotenberg recieved the Kai Siegbahn prize in 2009 for *the creation and development of the 'Electronic Structure Factory' end-station at the Advanced Light Source, which could legitimately be called the most useful ARPES end-station in the World.*

Since the present author was not involved in setting up the machine, no detailed description will be given, however most important information can be found online at http://www-bl7.lbl.gov/BL7/endstations/esf.html.

2.3 Special issues with the experimental set-up

UHV technique as well as photoemission spectroscopy specifically go together with many issues concerning the experimental set-up. This section will cover the explanations for many problems that can appear concerning the set up of an ARPES station that can perform Fermi-mapping.

2.3.1 Distortions in angular space

The understanding of possible and apparent distortions in angular space is fundamental for setting up ARPES machines. However, it is common to see the analyzer as a *black box*, which is probably resulting largely from the complexity of these problems; since

2. EXPERIMENTAL

the analyzer in an angular resolved lens mode projects the energy and the two angular dimensions onto three spatial dimensions, one has to deal with a quasi-six dimensional problem, where many factors have to be taken into account.

2.3.1.1 Curved and straight slits

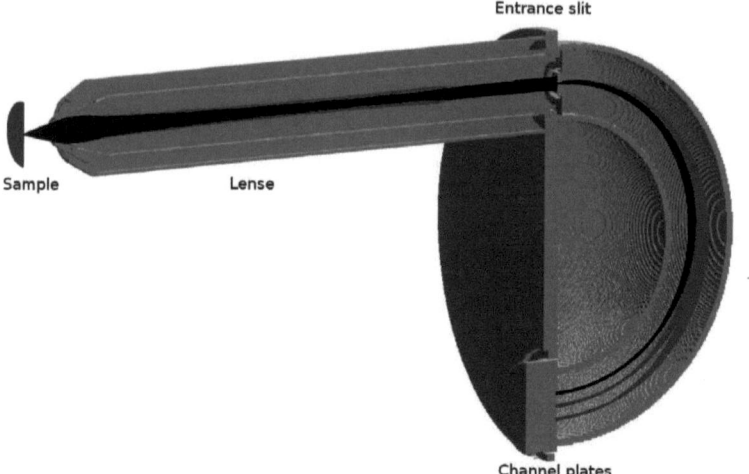

Figure 2.11: Schematic of the electron trajectories in the analyzer - The SIMION model was made by Thomas Braun [88]. Three different electron energies are marked by black, red and blue color.

In Fig. 2.11 one can see how the angularly and energetically resolved electrons hit the channel plates in the hemispherical analyzer. Since the electrical field within the hemisphere is spherically symmetric, a straight entrance slit will lead to a curved constant energy line in the angularly resolved photoemission intensity maps. This can best be seen in Fig. 2.12.

The red lines in the schematics on the left of this figure represent isoenergetic electrons. The blue lines represent the projection of the entrance slit on the channel plate plane. As one would expect, a straight entrance slit will lead to curved, constant energy lines on the channel plates. This distortion can easily be avoided by a curved

2.3 Special issues with the experimental set-up

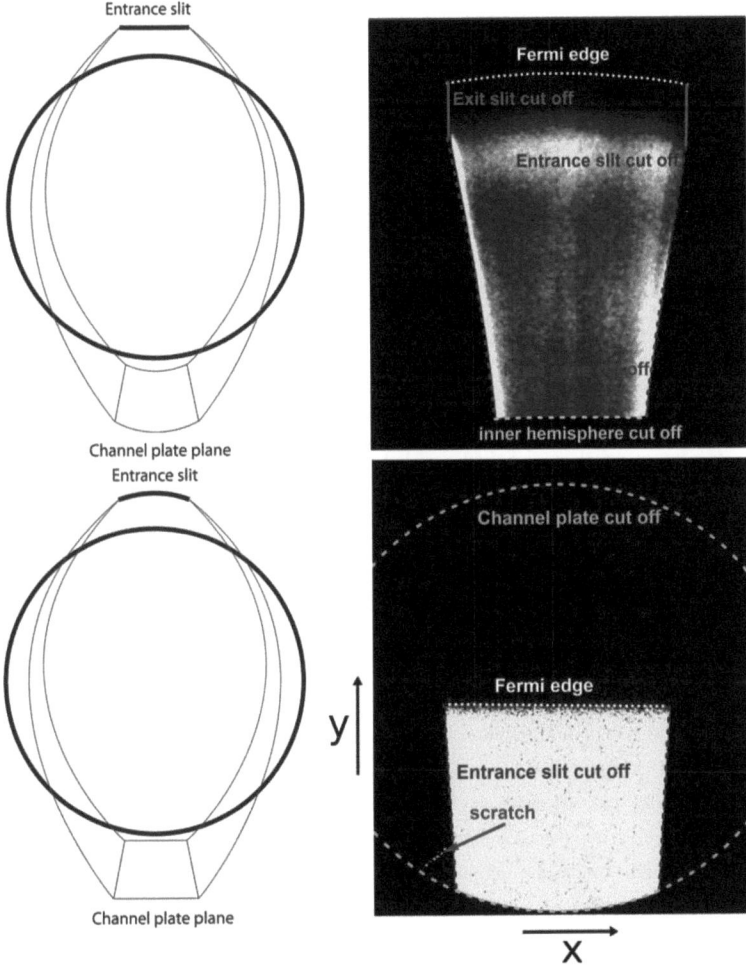

Figure 2.12: Schematic of slit-related changes in the photoemission spectra. - Left: schematics of the electron trajectories within the hemisphere for a straight (top) and a curved (bottom) slit. Right: Respective raw CCD-camera shots of the channel plates. x- and y-direction refer to θ angle and energy. Further information in the text.

2. EXPERIMENTAL

slit. However, as will be discussed later in this subsection, a curved slit has a somewhat distorting effect on our kind of valence band data sets.

On the right side of Fig. 2.12, shots from the CCD camera that reads out the signal from the channelplates for two spectra of polycrystalline metals close to the Fermi edge are shown. The upper spectrum is taken in WAM with a straight slit and a retarding ratio of 1 with the PHOIBOS100 analyzer. The y- and x-direction refer to the energy- and Θ-direction respectively. As one can see, in this spectrum the Fermi energy is clearly curved, since the electrons that enter the hemisphere further away from the middle trajectory are further away from the inner hemisphere and thus from the sides of the slit electrons with a slightly higher kinetic energy are passing through the hemisphere.

The photoemission signal on the channel plates is cut off on the sides. Depending on the retarding ratio and the lens mode these cut offs can be structured differently. In the upper panel on the right and left side, the photoemission signal close to the Fermi edge is cut off by two straight parallels from the exit slit of the hemisphere, as marked by the straight red lines. The exit slit cut offs are usually very sharp without any modification of the photoemission intensity close by. These cut offs usually play an important role when the measurements are performed in wide-angular acceptance modes with low retarding ratios, since this leads to a strong trapezoidal photoemission signal, which sometimes exceeds the diameter of the exit slit.

In contrast to the exit slit cut offs the nose cone cut offs usually show a strong photoemission signal directly before the cut. The cut offs originate in the lens system (usually at the nose cone), on which the electrons are scattered. This can only happen in lens modes with extraordinary high angular acceptance.

At the very bottom the cut off in the same subfigure (still Fig. 2.12, upper right image) results from the inner hemisphere. Depending on the kinetic energy of the electrons the spectrum is here either cut off by the hemisphere (upper spectrum) or by the channel plates (lower spectrum). Both are easy to distinguish since the channel plates are circularly shaped[1].

In the lower right part in Fig. 2.12 an image taken in low-angular dispersion mode with a curved slit and the PHOIBOS150 analyzer is shown. The Fermi edge is perfectly

---

[1]Deviations from the circular shape in Fig. 2.12 result from a small compression of the CCD-pictures.

2.3 Special issues with the experimental set-up

straight, and since the lens mode does not have a particularly wide angular acceptance, no scattered electrons from the lens system are detected. Instead, the spectrum is cut off by two rather sharp lines on the sides. These are the projection of the sides of the entrance slit on the channel plate plane. Those can be easily identified due to the trapezoidal shape that results from the hemispherical projection.

Additionally, a feature can be seen in this spectrum that results from a scratch on the channel plates (and is marked respectively). Our software allows us to automatically remove such effects.

As mentioned previously, for the valence band spectra primarily presented in this thesis a straight slit has been used since the curved energy line can be easily corrected, while the distortion that a curved slit creates in angular space is rather complicated to correct. Furthermore, as will be explained in subsection 2.3.1.2 even small magnetic fields in the chamber have a more distorting influence on the angular-space if a curved slit is used.

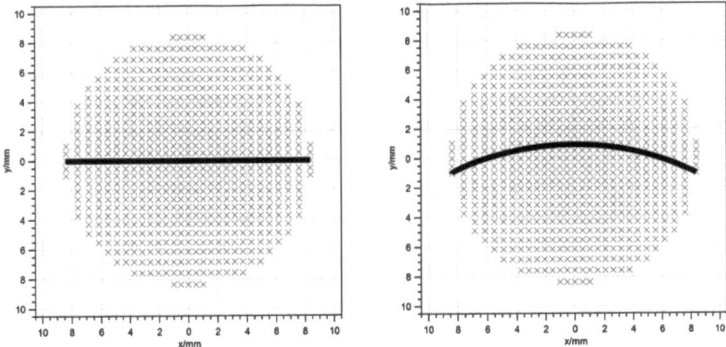

Figure 2.13: **Comparison of an entrance plane with a curved slit and one with a straight slit.** - The light grey crosses symbolize electrons coming from different angles from the sample. The x-coordinate stands in an isomorphic relation to the θ-angle, while the y-coordinate refers to the β-angle. As one can see, a straight slit will lead to spectra with a constant β-coordinate, while a curved slit will lead to spectra that different Θ coordinates will be coupled to different β coordinates.

How the distortion in angular space originates can easily be seen in Fig. 2.13. While a straight slit will let electrons through that have a Θ-independent β-coordinate,

2. EXPERIMENTAL

a curved slit will lead to a coupling of these two coordinates.

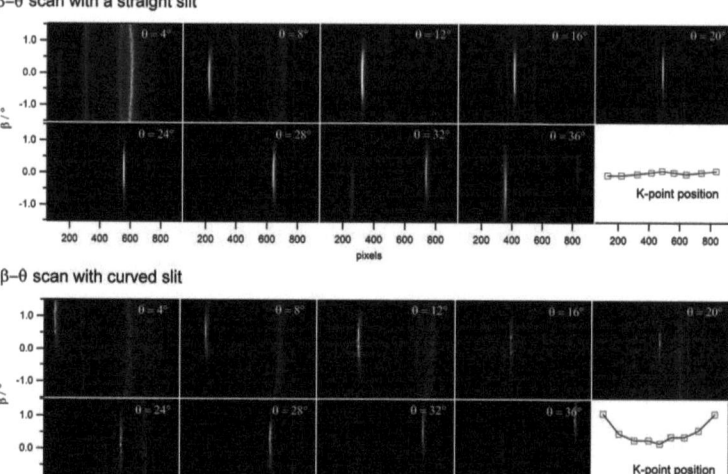

Figure 2.14: Constant energy cuts at the Fermi surface for different θ positions for graphene on Ru(0001) taken with a straight and a curved slit. - The raw data set has been taken at the ESF and no corrections have been made. One can clearly see the different trajectories of the K-point in the CCD-camera picture (In the lower right panel for both scans).

Graphene serves particularly well to demonstrate this distortion in angular space, since it has a particularly sharp feature at the K-point. By performing β scans for different θ angles (which we then named β-θ scan) we can track the movement of the K-point on the CCD-camera picture. This is demonstrated in Fig. 2.14, where constant energy cuts approximately 500meV below the Fermi level of graphene on gold on Ru(0001) are shown[1]. By tracking the bright feature at the K-point (see for both scans the lower right panel), one can see the projection of the slit in CCD-camera pixel vs. β-angle coordinates. The aberration from a strictly slit-curved function in the

---
[1]The data demonstrated are totally raw and thus for the spectra taken with a straight slit, the energy scale is curved. Consequently, the term "*constant* energy map" is not actually correct.

2.3 Special issues with the experimental set-up

respective graph results from the lens mode.

In conclusion, all valence band spectra should always be performed with straight slits to avoid distortions in k-space, while for core-level measurements a curved slit should be used, since core levels show no or only weak dispersion and thus no k-space distortion can be observed.

2.3.1.2 Magnetic fields

A big issue in the field of Fermi-mapping by photoemission is the presence of unwanted magnetic fields in the vacuum chambers. To protect photoemission stations from external magnetic fields usually μ-metal shielding is used. μ-metal is a nickel-iron-alloy (approximately 75% nickel, 15% iron and additional copper and molybdenum) with an extremely high magnetic permeability. However, the issue in the experimental set-up is usually not the μ-metal shielding itself, but rather the couplings of the separated μ-metal shells of each of the different parts of the experiment. In our case the magnetic coupling from the analyzer nose cone to the inner μ-metal shield with a μ-metal rim was the weakest part of the shield. Most magnetic field in the chamber disappeared, when the rim was additionally shielded by an extra μ-metal collar.

The lab machine, which is made out of stainless steal, has an inner 5 mm thick μ-metal shielding that was installed in March 2007, since first measurements with the machine showed strong distortions in angular space.

The BESSY endstation is made out of μ-metal, which makes further magnetic protection unnecessary. μ-metal is a suitable material for ARPES-stations due to the strong magnetic shielding, but at the same time it is softer than stainless steal, which can lead to unwanted deformations. Moreover, it can make a chamber useless if it becomes accidentally magnetized.

Since the major part of the electrons' trajectories is situated within the lens and analyzer hemisphere, the μ-metal shielding of these two parts is particularly important. Thus, the lens and the hemisphere are double-shielded by two layers of 1.5 mm thick μ-metal cladding. The μ-metal shielding within the lens will not only protect the lens from outer magnetic field, but it will also bundle the background magnetic field from the chamber, and thus the maximum rest field strength will be found within the lens. This is demonstrated in Fig. 2.15, where a curve of the magnetic field within the lens shows a dependence on the depth in the lens. As one can see, the highest magnetic

2. EXPERIMENTAL

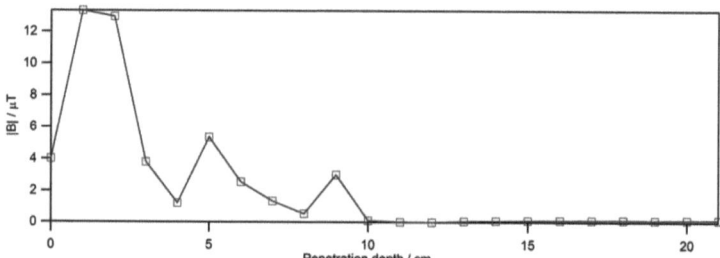

Figure 2.15: Magnetic field in the lens of a PHOIBOS225 analyzer by SPECS.
- With the kind permission of Sven Mähl

field is only 2cm behind the lens entrance. Here the electrons still have almost their original kinetic energy, and have not been accelerated or decelerated by the lens to reach the pass energy. Thus, one would expect a strong influence on magnetic field-induced distortions on changing the kinetic energy and only a weak effect from modifying the pass energy.

SPECS GmbH approves a magnetic field with a maximum of $2\mu T$ in the chamber for ARPES-measurements[1]. However, of course even small magnetic fields will influence the electron trajectories and while electrons with high kinetic energies will be influenced rather weakly by the rest magnetic field, slow electrons will be strongly affected. Thus, all SPECS analyzers have a tunable coil in the inner hemisphere to annihilate weak magnetic fields.

In a first approximation, one may visualize the effect of a magnetic field on angular space mapping by a closer inspection of the entrance-slit plane, as demonstrated in Fig. 2.16. The appearance of a field is here symbolized by a simple shift of the electron trajectories in the x-y-plane.

A perpendicular shift is slightly more complicated: In a first approximation it can be visualized by examining Fig. 2.7; a shift by a magnetic field can be approximated by a shift of the entrance plane, and thus will slightly scale the angular window. Moreover, since only at the entrance plane will electrons from the same exit angle and different energies have the same x-y-position, one would expect a slight angular dependence of

---
[1]The values of the magnetic field in Fig. 2.15 greatly exceed this value, since the measurements were taken in an *open* analyzer that was not mounted on any shielded chamber.

2.3 Special issues with the experimental set-up

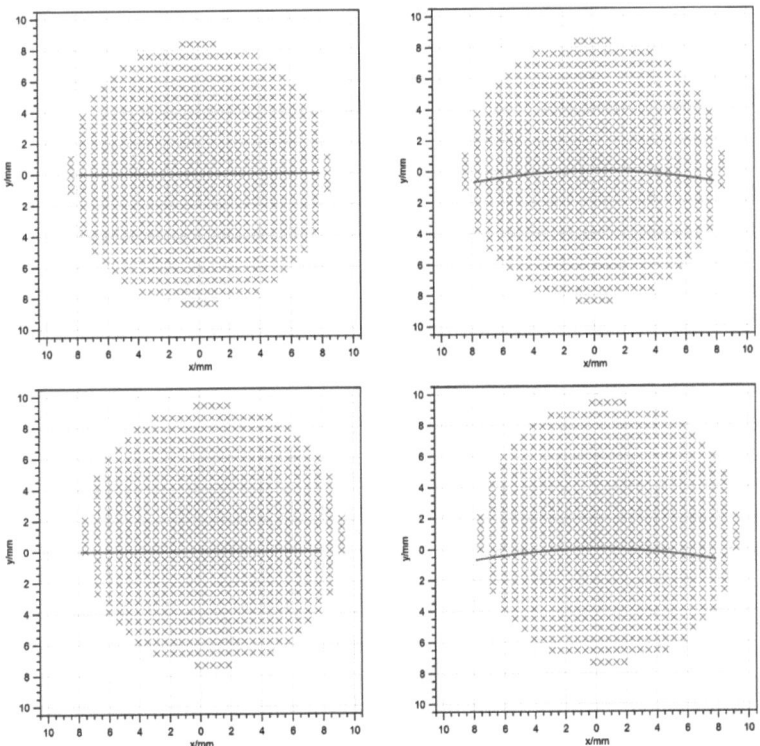

Figure 2.16: Schematic of the entrance plane with and without a field-induced shift. - In case of a shift of the electron trajectories (lower panels), the effect on the angular space with a straight slit (left panels) will be an asymmetry in the spectrum. In the case of a curved slit the effect is obviously more complex.

2. EXPERIMENTAL

the energies in the spectra. However, since this effect should be weak in comparison to the changes induced by the shift in the x-y-direction, it will not be further discussed here[1].

It is obvious that in the case of a straight slit one major effect will appear: the lens-induced distortion of the angular space will not be symmetric to the center. This is not a serious problem and can usually be eliminated by correcting the angular space digitally with the help of a grid that can be installed directly in front of the nose cone. As a consequence one will observe a distorted grid in the spectrum and can use this as a reference image for further correction.

Aside from the previous effect, in the case of a curved slit a far more complex distortion appears in the spectra, since then the asymmetry in the y-direction (β direction in angular space) crucially influences the angularly resolved spectra.

In Fig. 2.17, constant energy maps at the Dirac point of graphene on SiC in three β-Θ scans are shown, one with a maximum magnetic field within the lens of 16μT, one with 8μT and one with 2μT. The scans were taken in the lab chamber with the UVS300 He-lamp and the PHOIBOS150 analyzer by SPECS^{TM}, for which a maximum magnetic field of 2μT in the chamber is recommended by the manufacturer as mentioned previously. All measurements were performed in a completely identical experimental configuration. The only parameter that was modified was the current that was applied through a coil mounted around the analyzer lens flange to control the magnetic field in the chamber. With no current on the coil, the maximum magnetic field was 6μT in the chamber, which exceeds the nominal maximum value. However, unfortunately such a configuration with a permanently mounted coil should not be used to compensate the magnetic field in the chamber during *normal* measurements since the induced magnetic field will magnetize the μ-metal shielding in the long term.

As mentioned previously, a strongly asymmetric distortion as demonstrated here in this subsection was detected in an earlier version of the lab machine long before the present author started to work in the group. The setup consisted of the PHOIBOS150 analyzer, the lab vacuum chamber and an earlier version of the goniometer. BESSY was used as a photon source. Since only weak μ-metal shielding was installed and the analyzer was without straight slits, the distortion appeared both extremely strong

---
[1]The effect should be weak, since the electrons hit the entrance plane nearly perpendicular. This is demonstrated in Fig. 2.7.

2.3 Special issues with the experimental set-up

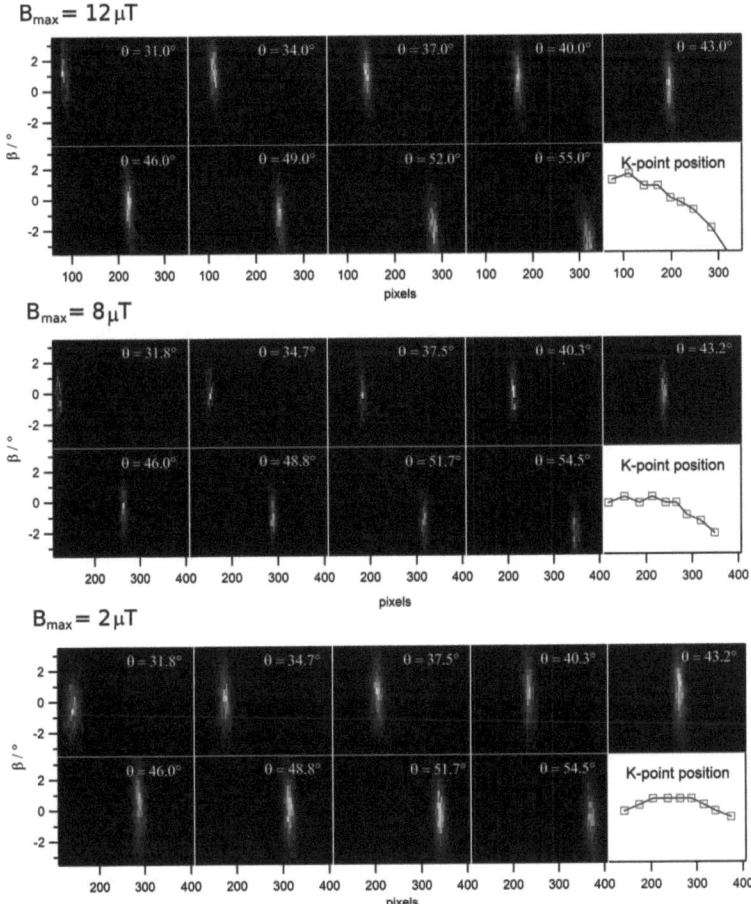

Figure 2.17: β-Θ **scans for a curved slit and different magnetic fields in the chamber** - The 2μT scan shows a nearly symmetric behavior of the K-point movement, while the 8 and 16μT scans strongly deviate from the symmetric form.

2. EXPERIMENTAL

and asymmetric. Since the reason for the distortion was not clear at that time, a long-lasting discussion with the manufacturer started as a consequence. The SPECS™ support had difficulties to figure out the exact reason for the distortion, because they were not used to the concept of β scans. At the same time, the lack of expertise concerning the functionality of a hemispherical analyzer and the lens system prevented the research group's finding the solution of the problem[1]. One hypothesis was that the analyzer has an intrinsic defect. Therefore, aside from experiments controlled by tuning the obvious parameters such as the kinetic energy of the electrons, the analyzer was once exchanged and once turned upside down to figure out the exact reason for the distortion in angular space.

These experiments did not lead to a solution, since every time the analyzer was turned or exchanged the magnetic coupling of the μ-metal shield of the lens and the μ-metal shield of the chamber became worse, and thus the effect became progressively stronger.

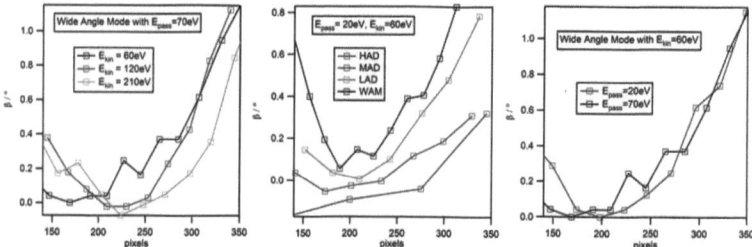

Figure 2.18: **Pixel-dependent K-point position for different kinetic energies, lens modes and pass energies.** - Higher kinetic energies move the extremum of the arc closer to the center. Lens modes with a lower angular acceptance limit the aperture of the arc. As one would expect, in agreement with Fig. 2.15, the modification of the pass energy has only a weak effect.

The hypothesis that the magnetic field combined with the curved slit is responsible for the distortion is particularly supported when taking a closer look at the dependencies of the K-point movement of the different analyzer parameters. This is best

---

[1] Although the theoretical possibility of a magnetic field as the cause was always under discussion; but magnetic field measurements in other ARPES chambers showed that much higher magnetic field than in our stations were very common and the effect of a curved slit was underestimated.

2.3 Special issues with the experimental set-up

demonstrated in Fig. 2.18. As one would expect, a strong dependency of the kinetic energy of the electrons on the K-point movement could be verified. Higher kinetic energies make the slit-projection symmetric, while low kinetic energies (as i.e. in the lab chamber) lead to a strong asymmetry in the observed curve. Furthermore, lens modes that cover a lower angular range logically reduce the effect. Also the weak dependency on the pass energy stands in perfect agreement with the hypothesis. Finally, no dependence on the iris aperture and the slit-size could be detected.

In summary it should be mentioned that we did not manage to lower the magnetic field down to $2\mu T$, however with straight slits such small rest magnetic field are not significantly distortive, while for core level measurements the β-θ distortion, as induced by a magnetic field and a curved slit, has no effect on the data.

2.3.2 Alignment

The following description of the alignment procedure assumes that the reader already has read the respective manuals, or is otherwise sufficiently skilled in controlling a photoemission station with a hemispherical analyzer with channel plates.

Schematic of front view of the sample with its fixed axes

Schematic of a missaligned sample. The sample should be at the position of the θ rotation axis.

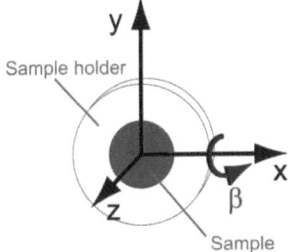
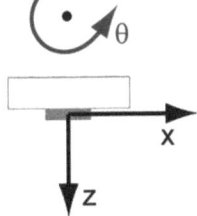

Figure 2.19: The alignment. - The axes that are important for the alignment are drawn in.

Since in contrast to most other ARPES stations the goniometer has six axes, the alignment of the system is a rather tricky issue. On the one hand the beam has to hit

2. EXPERIMENTAL

exactly the θ rotation axis of the chamber[1] (see Fig. 2.19) where the analyzer has also to be pointed. Furthermore, the sample has also to be directly at the rotation axis. For these reasons, the issue of a proper alignment is to find the z-position in which the sample surface is directly on the θ-axis, in order to have the beam best focused on the same position, where the focus of the analyzer lens should also be situated. Usually the first point presents the biggest issue.

Before the exact alignment procedure is explained, it should be mentioned that in all our chambers the x- and z-axes are sample-fixed. This means that changing the x-position will always move the sample left and right from the perspective of the sample. Thus, if the photon beam strikes the sample and one changes the x-position of the sample, the beam will roam over the sample without changing the alignment. In contrast, movement along the z-direction will change the alignment and thus this axis is crucial for the alignment of the goniometer.

For the alignment, the following axes, as shown in Fig. 2.5, have to be moved:

- The z-axis, which controls the movement of the sample perpendicular to its surface in the $\beta = 0$ position.

- The θ-axis.

- The β-axis.

- The y-, and the x-axes.

Beginning from scratch, the z-axis will be in a certain, arbitrary position. The θ- and β-axes should be moved to normal emission position[2], and the y- and x-axes should be moved in such way that the analyzer points to the middle of the sample. Before the sample will be moved in the correct position, one should assure that the channel plates of the analyzer are switched off. Then one should open the iris, put in a wide-open slit, and look through the alignment window of the analyzer. Now, one can move the goniometer in x- and y-direction until the analyzer points to the middle of the sample.

Subsequently, the analyzer and channel plates should be switched on. For alignment, one should use a spatially resolving lens mode (in case of low photon energies the *High*

---
[1] The θ rotation axis is fixed to the chamber, not to the sample.
[2] *Normal emission* position means the sample is pointing along the analyzer lens axis.

2.3 Special issues with the experimental set-up

Magnification 2 mode serves best). One can check if there is a weak photoemission signal from the sample by putting the channel plates on their maximum voltage, and shutting the photon flux on and off. Then the lamp, chamber alignment mechanism or beam line mirror (whatever is the device that controls the relative beam position to the chamber) should be moved in such a way that the photoemission intensity on the sample is maximized. At some point one will see the beam with the detector as a more or less thin line in the spectrum and then the beam should be moved in such way that it is situated in the middle of the CCD-camera picture so its intensity should be maximized. After the exact position of the beam on the CCD-camera picture has been written down, the goniometer should be moved around the θ-axis in such a way that the sample faces increasingly towards the beam.

This will induce a movement of the beam position on the CCD-camera picture (if the alignment is not perfect). At some point the beam will even move out of the picture. It should be noted how fast this happens (i.e. one can write down the θ position for every 100-CCD pixel movement of the beam).

At this point, the sample should be turned back into normal emission and the z-axis should be moved by a few centimeters in an arbitrary direction. Subsequently, the beam should be aligned again until it is at the same pixel position as before in normal emission. Afterwards, the θ axis should be moved again in the beam direction and it should be noted down again how many pixels by degree the beam moves. If the movement is now less than before, the alignment has improved, and thus the next iterative alignment step should be performed in the same z-direction as before. If the movement has increased, the alignment should be continued in the other z-direction.

After several iterative steps, one will find a z-position that allows the beam to stay in the middle of the CCD-picture until the θ-axis has moved so far that the analyzer cannot detect any photoelectrons since the framework of the goniometer is in the way. Further alignment can then be performed by turning the θ-axis in the opposite direction (away from the beam). However, this will sharpen the entrance angle of the photons on the surface and one has to decide *how* perfect the alignment should be. Due to temperature changes the alignment can never be perfect for a long time[1], but it can

---

[1] Usually a well-aligned chamber can be used for weeks without realignment. However, if one wants to make full Fermi-mapping, it can happen that he has to measure in goniometer position with a very sharp entrance angle of the beam. Here, temperature-dependent movements in the micrometer range can influence the measurement.

2. EXPERIMENTAL

easily be sufficient. Thus, the alignment only has to be perfect within the angular range in which measurements will be performed in the subsequent experiments.

Subsequently, the height has to be aligned by turning the β-flip. When during β-movement the beam moves out of the window, one can move the y-axis until the beam is in the middle again. When the β-axis can be moved all along the angular range that will be in use during the subsequent experiments without any recognizable change of the beam position, the y-axis is aligned.

After this procedure the alignment is done. However, after a new sample is in the goniometer the z-axis (*only(!!!)* the z-axis) should be realigned.

2.3.3 Ultra-high vacuum issues

To achieve ultra-high vacuum conditions, all chambers used for this thesis have rough-pumps to create a prevacuum in the 0.1mbar range. Such a vacuum is sufficient to get a turbo-molecular pump started, which can then pump the chamber down straight in the 10^{-9}mbar range without further vacuum improvement such as a bake-out or additional UHV-pumps.

Bake-outs are needed since the metal surfaces within the vacuum chamber are usually covered with thin water films. These water films permanently *evaporate* in the chamber. Since the vacuum in the volume tends towards thermal equilibrium with the surfaces and the majority of the water molecules stick on the surfaces, pumping a chamber down to the UHV range can take arbitrarily long[1]. For this reason the chambers were always heated up by heating tapes to a temperature over 100°C for about 36h. Overheating was prevented by controlling the temperature with several thermocouples, since many parts of the stations should no be heated to excessive temperatures (i.e. channel plates, electrical feed-throughs etc.). After such a bake-out cycle ultra-high vacuum conditions could be achieved.

TMPs are very efficient pumping systems, although their ability of pumping small molecules is very limited. However, hydrogen is usually quite an issue, since the metal UHV-chamber cannot prevent hydrogen from permanently entering the chamber. For this reason, titanium sublimation pumps were used; the evaporated titanium binds reactive molecules such as hydrogen, and then sticks on the surfaces of the chamber.

---

[1]Without a bake-out even weeks of pumping won't be enough to achieve pressures in the low 10^{-10}mbar range

2.3 Special issues with the experimental set-up

Additionally in all chambers ion pumps were used. These pumps work with an extremely high electrical field that ionizes the molecules. These molecules then also tend to react with the surfaces nearby.

At the ESF an additional extremely effective cryo-pump is used. Since molecules tend to stick on cold surfaces a small cooled part guarantees very quick pumping.

2. EXPERIMENTAL

3
Graphene on ruthenium

3.1 Introduction

As previously described in Chapter 1, the extraordinary properties of graphene render it a promising candidate for future electronic devices [34]. Most of the interesting properties of this material derive exclusively from the equivalence of the quasiparticles to massless Dirac fermions, and the conical shape of the π and π^* bands which cross only at two points (K and K') of the Brillouin zone.

The lack of an energy gap near these crossings (at the Dirac energy E_D) limits the potential for applications, and the preparation of graphene-based systems with a gap is an important step towards future graphene device engineering applications. The simplest mechanism for opening a gap is by breaking the symmetry of the two carbon sublattices which protects this gap [89]. This is the case in armchair nanoribbons[1] [91], in biased graphene bilayers [92, 93] or by breaking the sublattice symmetry by bonding graphene (or bilayer graphene) to a substrate [94].

As discussed in chapter 1, epitaxial graphene on SiC(0001) presents an anomalous bandstructure near E_D which has alternately been interpreted as due to many-body renormalization (known as "kinking") of the bands due to electron-plasmon scattering [25, 50, 95, 96, 97, 98, 99] or else due to symmetry breaking due to substrate interaction as outlined above [58, 100]. Such behavior should be distinguished by the observation of the presence or absence of an energy gap, but the interpretation of the data has been

---
[1] The sublattice symmetry breaking in carbon nanoribbons happens due to different site energies of the carbon atoms at the edge. See Ezawa [90] and references therein.

3. GRAPHENE ON RUTHENIUM

controversial [59, 60]. A gap-like spectrum in disordered graphene has been demonstrated [101], so it is important that the number of defects be minimized in order to observe the intrinsic spectral function.

The Ru(0001) surface is particularly suitable for the study of graphene's interaction with a substrate for several reasons: graphene layers of extraordinary good quality can be easily grown [102, 103, 104], the second layer reveals a structure that is similar to free-standing graphene [66], and intercalation of noble metals can easily be performed, as will be described later in this chapter.

The data presented here serves to show that in a single system – graphene grown on the Ru(0001) surface – the behavior of the graphene bands near E_D can be selected between extremes, from exhibiting the specific many body kinks to a situation where a gap appears, by controlling the structure of the graphene-substrate interface. When the first layer below graphene is a graphene-like "buffer layer", the spectrum exhibits kinks due to many-body interactions, but when this buffer layer is replaced with a layer of Au atoms, the spectrum changes to an unmistakably gapped one. This constitutes the first direct observation of a gap at the Dirac point in a sample without the complication of small domain size or high defect density[1] [59, 60].

3.2 Apparatus

The ARPES experiments were performed at the Electronic Structure Factory at beam line 7 at the Advanced Light Source of Lawrence Berkeley National Lab (see subsection 2.2.3) using 95eV photon energy for all valence band spectra. The photoemission intensity data sets over the energy-momentum space (E,k_x,k_y) were collected with a Scienta R4000 energy analyzer with samples on a liquid He-cooled, 6-axis goniometer at $T = 20$K. The energy/momentum resolutions were 30 meV/0.01Å^{-1}. The base pressure during measurements was $< 7 \times 10^{-11}$ mbar.

3.3 Preparation

Before the graphene formation the ruthenium crystal was flashed several times at approximately 2000°C and annealed at 1000°C under an oxygen atmosphere of 10^{-7}mbar

---
[1] Some part of the data and conclusions presented here have been published in New Journal of Physics in March 2010 [57].

3.3 Preparation

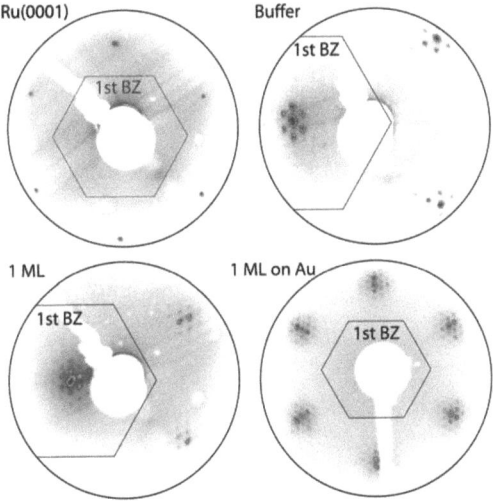

Figure 3.1: LEED images of different Ru(0001)-based systems. - Upper left panel: The clean Ru(0001) surface shows sharp LEED spots at 100eV electron energy. First BZ is drawn in. Upper right panel: The first carbon layer (here referred to as the buffer layer) shows similarly hexagonal LEED-spots. The lattice missmatch of the Ru(0001) surface and the buffer layer results in hexagonal satellite spots around the main spots. The LEED image has been taken with 60eV electron energies, since at higher electron energies the satellite spots show strongly reduced intensity. Lower left panel: LEED image of another graphene layer atop the buffer layer taken at 60eV. The LEED spots in the second BZ are much weaker. Although the relative intensity of the satellites increased. Lower right panel: LEED image taken at 100eV electron energy of a graphene layer over intercalated gold. The peaks in the second BZ are strong and show many satellites.

3. GRAPHENE ON RUTHENIUM

until no C1s core level photoemission peaks could be observed. The graphene layer was prepared using the carbon segregation method [66, 102]. In a small temperature window around 600°C the carbon atoms that are dissolved in the crystal segregate to the surface. Thus, to create graphene layers the sample was annealed to 1000°C and then cooled down slowly within 10min (for a monolayer), 30min (for a bilayer), or 90min (for three layers). In Fig. 3.1 one can see LEED images of the clean Ru(0001) surface, the buffer layer atop, one graphene monolayer and one monolayer on intercalated gold. The LEED images prove that the graphene layers grown by this method exhibit the same hexagonal lattice orientation as the Ru(0001) surface.

Other studies demonstrate the high quality of such graphene overlayers[102, 103]. The initial graphene layer on the Ru(0001) surface exhibits a hexagonal superstructure with a periodicity of 30Å [105], which is attributed to the lattice mismatch of about 10% between graphene (lattice constant $a = 2.46$Å) and the Ru(0001) substrate ($a = 2.706$Å) [105]. Ab initio calculations suggest that this mismatch is accommodated by a pronounced rippling resulting in the position-dependent strength of interaction with the substrate [39, 106], but experiments show partially contradictory results [102, 103]. The initial graphene layer, while metallic, does not show a clear π band crossing at E_F [107], while the subsequent one and two layers show mono- and bilayer graphene-like bandstructures [66], respectively. We therefore call the initial layer the buffer layer (in analogy with the situation on SiC(0001)) and subsequent layers the first, second etc. graphene layer.

Several times after recleaning the ruthenium surface it was found that no further graphene layer growth was possible, since not enough carbon was dissolved in the crystal. To *upload* the ruthenium crystal with new carbon, it was annealed to 1000°C under an ethylene atmosphere of 10^{-6}mbar for 30min. This resulted in carbon being adsorb into the crystal.

For the experiments presented in subsection 3.4.2 Au was intercalated under the graphene layer. This was accomplished by depositing a thick Au film ($>$ 3 ML) on top of the buffer layer, followed by light annealing of 600°C, after which most of the gold evaporates leaving a single layer intercalated underneath the buffer layer. The buffer layer was then transformed into a true graphene layer with sharp π bands and a clear Fermi surface (see Fig. 3.2).

3.4 Results

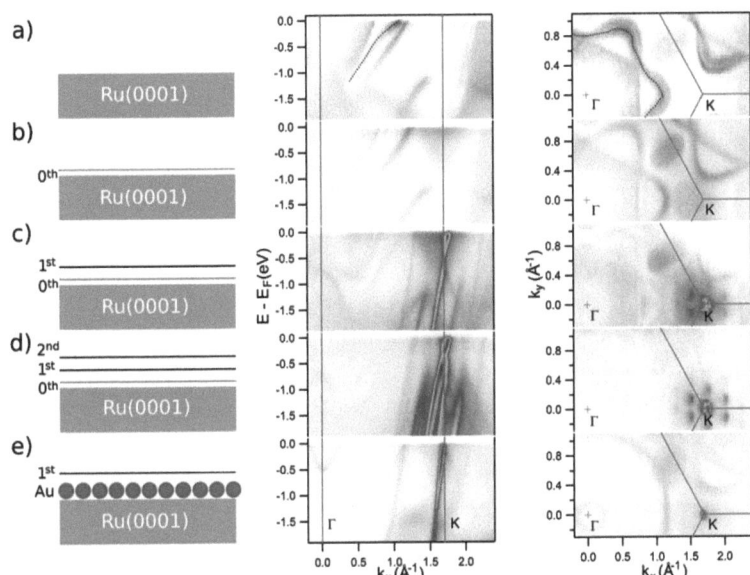

Figure 3.2: **Overview of the photoemission intensity maps of the measured systems** - Left column: schematics of the systems; namely the clean Ru(0001) surface (a), the graphene buffer layer on the surface (b), a quasi free-standing graphene layer on top of this (c), an additional second graphene layer atop (d), and the graphene/gold/Ru(0001) interlayer system (e). Middle column: band maps along Γ-K direction of the respective systems measured with 95 eV photon energy. High symmetry points are drawn in. Right column: respective Fermi surfaces. BZs are drawn in.

Fig. 3.2 compares the different systems that have been measured. The spectral function at 95 eV photon energy of clean Ru(0001) in subfigure a) stands in good agreement with previous studies [66, 83, 108], however it shows a Ru(0001) surface state as marked by the dashed line in the band map and the respective Fermi surface. That this band belongs to a surface state is clear due to its relative strength for clean Ru(0001) surface and its absence in the case of a covered surface[1].

---
[1] To clarify the nature of the state, we have performed live ARPES measurements, while covering

3. GRAPHENE ON RUTHENIUM

The spectral function of the buffer layer is characterized by the absence of sharp graphene π-bands in the vicinity of the Fermi level, in agreement with previous studies [17, 66, 107] that show that the Dirac energy[1] E_D in the buffer layer on ruthenium is shifted by 2.6eV to higher binding energies compared to those for pure graphite. This can be attributed to the strong interaction between this layer and the ruthenium substrate leading to a strong hybridization between the graphene π and Ru 4d valence band states[17, 39, 66, 107]. Such interaction leads to the formation of an unusual "cloudy" structure around the K-point, which is clearly visible in the band map and the Fermi surface. Although the interpretation of this cloudy feature as an energy band is not straightforward, we can take its general shape to indicate an electron pocket derived from the graphene π-band, but heavily modified by strong interactions of the graphene monolayer on ruthenium, and inhomogeneously broadened due to the spatially-varying interaction strength.

The formation of the first graphene monolayer (on top of the buffer layer) on Ru(0001) leads to dramatic changes in the electronic structure of the system. We now observe a linearly-dispersing π-band around the K-point which crosses the Fermi level, reflecting the massless behavior of electron carriers in the graphene layer. There is no apparent interaction between the bands in the two layers. The graphene layer, similarly to the graphene/SiC system [25, 50, 81], is n-doped with the position of the Dirac energy E_D at 0.5eV binding energy (BE) below the Fermi level E_F. We conclude that both layers are doped by charge transfer from the substrate, but only the upper layer's bands strongly resemble pure graphene in accordance with Sutter et al. [66].

Another interesting feature are the graphene *satellite* bands around the K-point that one can see in Fig. 3.2 c) that will be discussed in subsection 3.5.3.

As one would expect the spectral function of the graphene bilayer close to E_F is likewise similar to the graphene bilayer on SiC [25, 93, 109]. Instead of one π-band now two π-bands with a non-linear dispersion at the K-point appear in the photoemission spectra.

Previous studies of graphene on Ni(111), [110, 111] show that upon intercalation with noble metals, the π-states in graphene recover nearly the same band structure as

---

the Ru(0001) surface with hydrogen (data not shown in this thesis).

[1] The strong hybridization of the graphene π-bands with the Ru d-bands actually makes the term *Dirac energy* invalid. But since the respective *would be* Dirac energy level still has fundamental meaning in terms of doping, it will be used in this context.

pure graphene, exhibiting the linear dispersion of the π-bands in the vicinity of the Fermi level. This reflects the weakening of the interaction between the graphene layer and the substrate. Fig. 3.2 e) shows the electronic structure as well as a photoemission intensity map at the Fermi level of the 1ML graphene/Au/Ru(0001) system. Now, the buffer layer is transformed into a true graphene layer with sharp π bands and a clear Fermi surface.

3.4.1 Graphene layers of different thickness on ruthenium

Fig. 3.3 compares the spectral function of the buffer layer, and mono- and bilayer graphene on the Ru(0001) surface. As one can see all three Fermi surfaces (upper left panel in each column) do not show a homogeneous intensity distribution where the bands cross the Fermi surface. Instead the intensity on the left is much lower, or even vanishing in the case of the monolayer, than on the right in the Fermi maps. This can best be seen in the band maps along the Γ-K direction (panels below the Fermi maps), where in all three cases only one band reveals strong photoemission intensity. This effect was explained by an interference effect of the emitted electrons by Shirley et al. [112]. Since this interference effect strongly depends on the equality of the wave functions of the electrons in the two sublattices, the relative photoemission intensity of the two bands along the Γ-K direction can be seen as a direct measure of the strength of the symmetry breaking between the sublattices [81].

For the buffer layer, the graphene π-bands are only very weakly visible for reasons discussed above. However, the position of the bands let us estimate the doping of the buffer layer to 2.2($\pm$0.3)eV, which slightly disagrees with the value of 2.6eV measured by Sutter et al. via π-band shift at the Γ-point [66].

As shown in the lower left panel of Fig. 3.3, the photoemission intensity profile along the Γ-K line at $E = E_F$ has been fitted with two Lorentzian functions. Since the line width should primarily be determined by the natural line-width close to the Fermi-level, Gauß-folding was neglected. Two lorentz fits are necessary since along the Γ-K line around the K-point the π-band with decreased photoemission signal shows a rather strong intensity. The intensity ratio of the two peaks in the photoemission intensity profile, after subtraction of the background, is roughly 1 : 2, which would go together with a band gap of 5eV according to the model of Jones [89], as it was expanded by Bostwick et al. [25]. However, this model is based on a Hamiltonian of

3. GRAPHENE ON RUTHENIUM

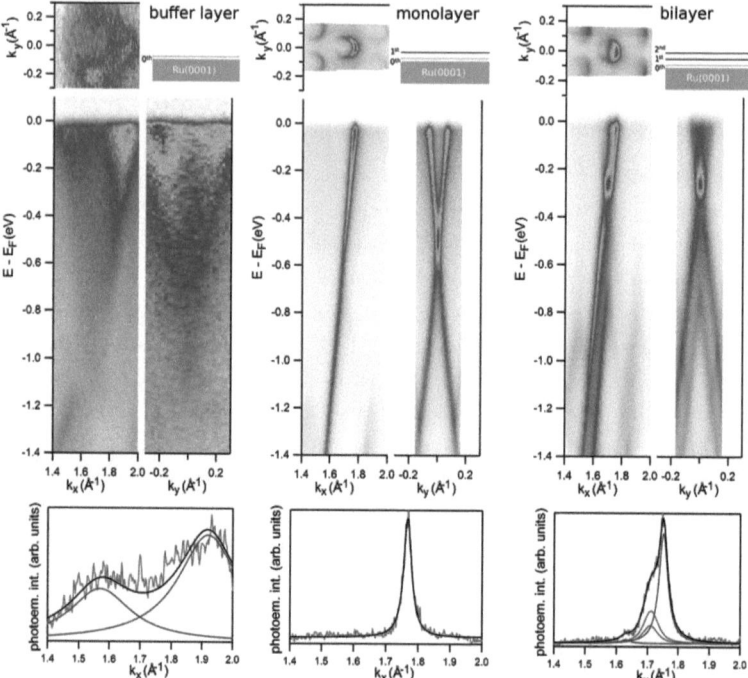

Figure 3.3: **Spectral functions and photoemission intensity profiles at the Fermi surface along Γ-K line around the K-point of graphene layers of varying thickness on Ru(0001)** - Upper panels: photoemission intensity maps around the K-point for the buffer layer (left column), monolayer (middle column) and bilayer graphene (right column) on ruthenium. Lower panels: Respective photoemission intensity profiles and fits at $k_y = 0$ and $E = E_F$.

3.4 Results

graphene around the K-point as presented in equation 1.14 in section 1.1.1.2, which is based on the assumption that the electrons still are close to Dirac-behavior. Naturally it is questionable if this approximation is valid in case of such a strong symmetry breaking. Moreover, due to the strong hybridization of the graphene π-bands with the Ru $4d$ states a discussion concerning the exact size of a gap might be difficult, since the hybridization might directly influence it. Finally, the hybridization probably leads to a change of cross sections within the hybridized bands[1] and thus might induce photoemission intensity variations that make the upper approach for an estimation of the gap size impossible.

For the monolayer the situation is different and no weak-intensity band is observed. Instead, the photoemission intensity map shows all the characteristics of quasi-particle interactions as previously found for graphene on SiC [50]. This quasi free-standing graphene layer is doped by 500meV, which is also close to the doping of 450meV as observed for SiC [50].

In the case of bilayer graphene, the fit has been performed with four peaks as shown in Fig. 3.3, right, lower panel. The inner two peaks result from a band that has its maximum above the Fermi level. Therefore, these peaks have been fitted with Gauss-folded Lorentz-functions due to the stronger broadening. Since the second graphene layer is stacked in such manner that the carbon atom of one sublattice is always above a hole, while the other is above an atom, the bilayer reveals a gap opening due to a strong symmetry breaking that can also be quantified by the relative intensity of the weaker band [25]. In our case the intensity ratio for the two outer peaks in the profile in Fig. 3.3) is 0.03, which relates to a gap-size of 200($\pm$50)meV and thus agrees with the directly measured value of 180($\pm$30)meV.

For SiC it has been shown that the doping-gradient between the two layers leads to strong modifications for the electronic structure of bilayer graphene in comparison to free-standing non-doped graphene [93]. Since the substrate-induced doping of graphene on Ru(0001) is rather similar to the doping of graphene on SiC, the bilayer situation is also similar.

Another remarkable feature are the satellite bands of the graphene π-states shown in Fig. 3.2 for mono- and bilayer graphene. The satellites reveal the same structure as the π-bands but are shifted by 0.23Å^{-1} in a hexagonal structure. The hexagon of the

---

[1] As it is the case for graphene on Ni(111) (see chapter 5).

3. GRAPHENE ON RUTHENIUM

satellites is 30° rotated with respect to the hexagon of the original graphene π-bands. A similar satellite structure has been observed for graphene on SiC and interpreted as a low-energy electron diffraction pattern from the buffer layer [25, 50]. This will be further discussed in section 3.5.

3.4.2 Energy gap formation in graphene on ruthenium by control of the interface

As mentioned previously, gold-intercalation underneath the buffer layer has been performed. We estimate, based on core-level intensity measurements (shown in Fig. 3.6 and discussed in subsection 3.5.1) that the Au layer thickness is about 1 ML, independent of the pre-deposited Au layer thickness. The same thickness of intercalated gold was obtained for the graphene/Au/Ni(111) system [113]. The spectral function around the Γ-point still reveals weak photoemission intensity from the Au surface state, indicating residual gold islands on top of the graphene layer [113]. Nevertheless the intensity of the graphene π bands and the relative weakness of the gold surface state proves that most Au has intercalated.

The intercalation of Au underneath the buffer layer on Ru(0001) leads to an energy shift of the π-band of the graphene layer to lower binding energies compared to the buffer layer, and mono- and bilayer graphene on Ru(0001). Charge transfer from the graphene layer to the substrate is particularly weak, as can best be seen by the fact that the Fermi surface appears to be pointlike in Fig. 3.2 e), which indicates zero or nearly zero doping of the graphene layer. However, a close look at the dispersion of the π-states around the Fermi level in Fig. 3.4 (left, lower panel) reveals that the Fermi surface is not really pointlike, since the Dirac-point is $\sim$ 150 meV above E_F, consistent with a small p-doping of the graphene layer in the 1ML graphene/Au/Ru(0001) system. A similar p-doping of graphene was recently observed in the case of deposition and annealing of a thin gold layer on the graphene monolayer on SiC(0001) [114].

In order to study the shape of the bands around E_D, it is necessary to push the band crossing below the Fermi level by n-type doping. This was done by controlled deposition of potassium atoms on top of the 1ML graphene/Au/Ru(0001) system [93], upon which the π-band and the Dirac-point shift rigidly to higher binding energies with increasing K dosage. Surprisingly, with increasing doping a clear energy gap for π-states becomes visible at E_D (Fig. 3.4; best visible in the lower right panel).

3.4 Results

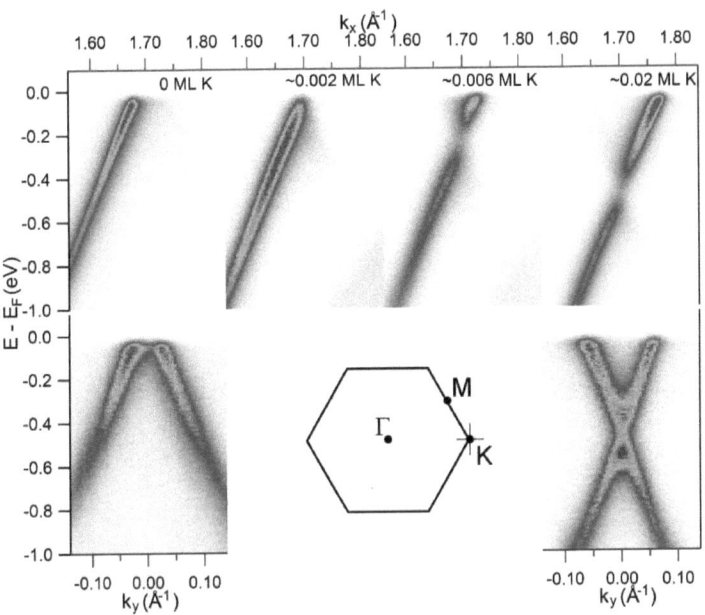

Figure 3.4: Doping scan for K on graphene on Au on Ru(0001). - A series of photoemission intensity maps around the K-point of the Brillouin zone of the 1ML graphene/Au/Ru(0001) system for clean (left column), and progressive doping with potassium. The upper and lower rows are taken along two orthogonal directions in the reciprocal space as indicated by the red and black line at the K point in the Brillouin zone (inset).

3. GRAPHENE ON RUTHENIUM

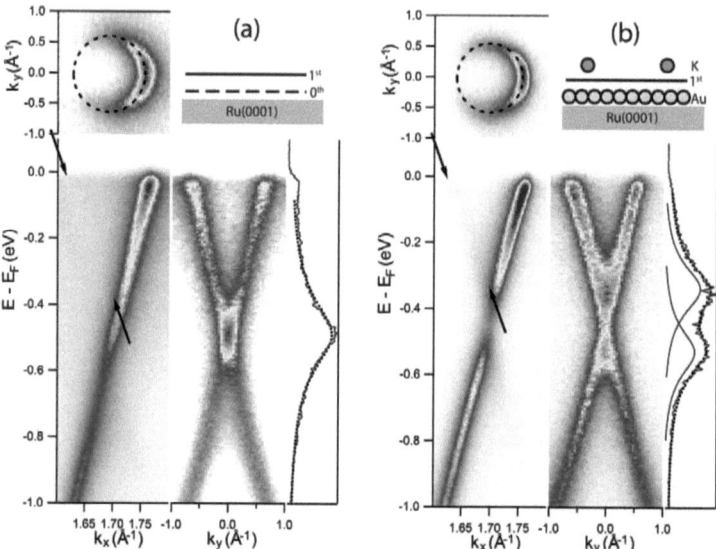

Figure 3.5: Comparison of the two quasi free-standing 1 monolayer graphene systems. - Comparison of the spectral function of one monolayer graphene on (a) buffer layer or (b) one monolayer gold atoms on Ru(0001). Three cuts of the spectral function are shown along (upper left) the Fermi surface vs (k_x, k_y) (lower left, right) the bandstructure along the two orthogonal cuts indicated in the inset to Fig. 3.4. For both systems the photoemission intensity along the K-point is plotted and fitted with one Voigt-peak for the buffer + first, and with two Voigt-peaks for the Au + first graphene layers on Ru(0001).

Comparison of the two one-monolayer-graphene-on-ruthenium band structures is shown in Fig. 3.5. From the conical shape of the π^* bands, the K point spectrum (and hence the Dirac crossing energies E_D) can be unambiguously determined. For the graphene/buffer layer system [Fig. 3.5 (a)] we find that, within the limits of our experimental resolution, 30 meV, the band structure at the K-point exhibits no gap. Instead it has a weak kink around E_D, consistent with the influence of electron-plasmon coupling as reported for graphene on SiC [50, 81] and predicted for free-standing graphene [96, 97, 99]. The energy distribution curve (EDC) at the K-point [see right panel of Fig. 3.5(a)] can be represented by a single Voigt peak with a full width at half maximum (FWHM) of 100meV. In contrast, n-doped graphene on Au on Ru shows the clear formation of an energy gap at E_D; an analysis of the EDC at the K-point shows that it requires two peaks (of FWHM 100 meV) to model the data, demonstrating that the band structure of this graphene layer exhibits a gap of about 200 ± 30 meV gap. Since we acquired the spectra in Fig. 3.5 by a fine sampling of the entire two-dimensional momentum range, we can be sure that the presented band structure cuts passed precisely through the K-points, and therefore the observed gaps cannot be due to sample misalignment.

3.5 Discussion

3.5.1 The thickness of the gold layer

Before the origin of the band gap will be discussed, it should be clarified that the amount of gold that has intercalated is indeed only one monolayer. This can best be seen in the core level spectra in Fig. 3.6 a), where the C1s (at 284 eV) and the Ru3d (at 284 and 280 eV) core levels are shown for the clean surface, the buffer layer, and the two graphene monolayer systems discussed here. The spectra have been normalized to the Ru$3d3/2$ peak at 280 eV. If we consider the mean free path of gold equal to the mean free path of carbon as a first approximation, we would expect the 284 eV peak of the graphene/Au/Ru(0001) spectrum being roughly the respective peak of the graphene/buffer/Ru(0001) system subtracted by the peak of the buffer/Ru(0001) system, since no carbon photoemission signal would be expected other than from the graphene layer. This is roughly the case. If more than one monolayer had intercalated

3. GRAPHENE ON RUTHENIUM

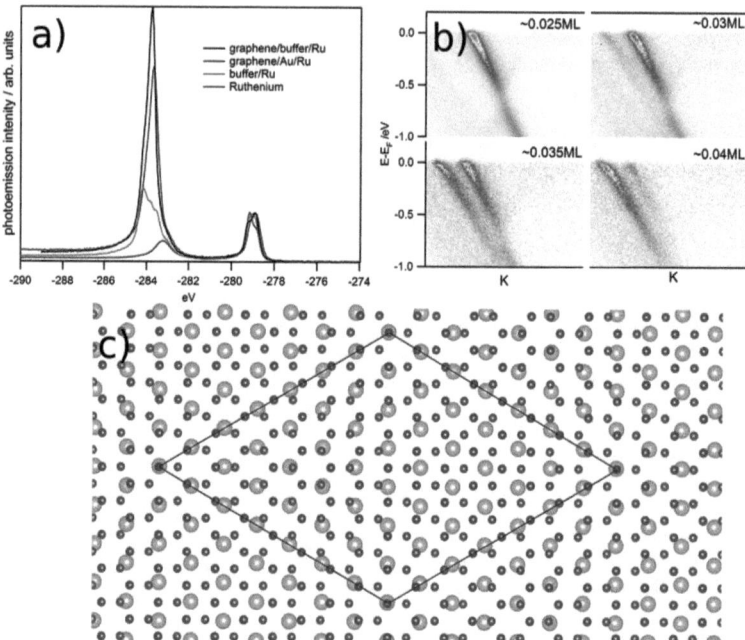

Figure 3.6: **Graphene on intercalated gold on Ru(0001).** - a) Core level scans around 280 eV binding energy of Ru(0001), the buffer layer, the monolayer, and one monolayer on intercalated gold. b) Spectral function of graphene on gold on Ru(0001) around the K-point in Γ-K direction for higher potassium coverage then shown in previous figures. One can see clearly the phase transition via the appearance of a second π-band and the disappearance of the original one with increasing potassium coverage. c) Artist view of one monolayer of graphene on a Au(111) layer with a lattice constant of 2.81Å. The unit cell of the superstructure is drawn in.

3.5 Discussion

underneath the graphene layer, the 284 eV peak would be significantly higher than the graphene/buffer/Ru(0001) peak.

3.5.2 The origin of the band gap

The band gap opening at the K-point in the monolayer graphene on one monolayer of gold on the Ru(0001) surface seems particularly surprising, since theoretical studies actually predict that graphene on gold remains gapless and even experimental studies suggest that no gap opens (for both see [115])[1]. Before a possible explanation is given for this apparent problem, the mechanisms that might cause a band gap should be discussed.

- A gap opening could result from the formation of a potassium superstructure on the graphene layer, as reported by Pivetta et al.[116]. In our case this can be excluded, since the formation of such a superstructure is a rather sudden phase transition that can easily be determined in our data sets, via an apparent change in the band structure. Our measurements begin to show such phase transition with higher potassium coverage, as demonstrated in Fig. 3.6 b).

- Other potassium-induced mechanisms that lead to the opening of the band gap are also rather unlikely, since the band gap does not increase with higher potassium coverage.

- A mechanism of gap opening at the K-point that is *only* caused by a hybridization of the p_z bands with gold states is unlikely. Theoretical calculations of graphene on nickel show that even in hybridized graphene systems the non-gapped spectral function should be preserved at the K-point [42]. As long as the symmetry of the graphene sublattices is preserved, there is no reason, why a gap should open (see Chapter 1). However, a hybridization with gold states is very likely[2]. As shown in chapter 1 subsection 1.1.1.2 the π and π^* bands must touch, as long as the sublattice symmetry is preserved. However, a hybridization-induced band gap

---

[1] Ref. [115] is particularly interesting, since the gap opening of K-doped graphene on Au is attributed to the K-atoms. However, the data of Varykhalov et. al. shows also an unusual foto emission intensity spectrum at the K-point, which the might result from bad alignment of the sample.

[2] Our data does not show any hybridized states, but of course the hybridization can happen also above the Fermi-level.

3. GRAPHENE ON RUTHENIUM

could likely change with varying doping since the gold bands would not shift in the same manner, meaning that the energetic position of the Dirac-point could change in relation to the hybridizing Au-state, as observed for graphene on Ni(111) [67]. No such effect was observed.

Thus we propose that the band gap in the 1ML graphene/Au/Ru(0001) system results from a symmetry-breaking of the two carbon sub-lattices in the graphene layer. This results in a weak breaking of the chiral symmetry, inducing a weak but finite intensity of the left band along the Γ-K direction shown between the two arrows in Fig. 3.5(b) [25, 112, 117]. The ratio of the left to right band intensities in Fig. 3(b) is about 35, which agrees with theoretical predictions for the size of a gap of 200meV [25]. Within our statistics, there is no equivalent observable intensity for the graphene/buffer layer/Ru system [between arrows in Fig. 3.5(a)], consistent with lack of a gap at E_D.

As mentioned previously, the appearance of the gap in gold-intercalated graphene still seems surprising. Calculations indicate a weak bond between noble metals and graphene without a notable gap [94]. While a detailed explanation of why gold intercalation opens a gap will depend on the exact microscopic structure of the interface, which is outside the scope of this work, we can speculate that the incommensuration between graphene and Au lattice constants (2.46 and 2.81Å, respectively), neglected in the calculations [94], plays a role. Depending on the exact length scales, the K and K' points of the graphene can be coupled, amounting to a breaking of the sublattice symmetry in real space that protects the Dirac point from opening a gap. The model of the respective stacking is shown in Fig. 3.6 c), which shows a simple speculative model of the graphene-gold interface. Since each sublattice in the picture is shown in a different color, one can see that the potential felt by one graphene sublattice from the gold layer will always differ from the potential felt by the other sublattice. It is crucial to understand that this is the case in either way the graphene lies atop the gold, as long as the orientation is preserved, which is proven by the LEED images (Fig. 3.1) and band maps (Fig. 3.2) down to a rotation of 0.5°.

3.5.3 The origin of the satellites

In contrast to the ideas presented in the previous subsection, the strength of the photoemission satellite bands is greatly reduced or even vanishing for interfacial Au compared

3.5 Discussion

to the buffer layer. According to Bostwick *et al.* [25] this would be consistent with a related weak potential associated with the Au incommensuration[1]. Thus, from this point of view it seems that these two observations are contradictory.

However, the satellite spots arise from an electron diffraction but this actually does not contradict the line of argument of the weak interaction between the graphene layer and the buffer layer, since the diffraction spots cannot reveal from simply diffracted electrons from the buffer layer, but rather from the diffraction of the superstructure from buffer layer and the substrate. Before this will be explained in detail, three strong arguments against the idea that the strength of the satellite bands can be seen as a measure of the strength of the interaction between the graphene layer and buffer layer will be given:

- Graphene on the buffer layer on Ru(0001) and graphene on the buffer layer on SiC(0001) are systems that can be very well compared. These systems only differ by the substrate underneath the buffer layer, which should not induce much change in the band structure in the graphene monolayer in either case [50, 66]. The only apparent difference lies in the orientation and distance of the satellite peaks. In detail, the orientation of the satellites is rotated by $30°$ in relation to the graphene in the case of Ru(0001) (see Fig. 3.2) and is not rotated in the case of SiC [50]. However, the only structure in real space that is also rotated by $30°$ relative to the graphene in these systems is the substrate: while graphene on ruthenium has the same orientation as the substrate, graphene on SiC has an orientation that is rotated by $30°$. Note that the orientation of the satellites is rotated on ruthenium, while the substrate is not rotated, and in SiC the opposite is the case [56]. This naturally is the case, if the spots arise from the buffer/substrate superstructure, as will be explained below, but is puzzling if the satellites arise from the interaction between the graphene and the buffer layer.

- The distance between the satellite spots and the main spot accounts for roughly 0.3Å^{-1} in the case of ruthenium and about 0.5Å^{-1} in the case of SiC[2]. However, the buffer layer will have roughly the same lattice constant in both systems and

---
[1] And the same theory has been used by the present author [57]. However, here I will show that this idea was based on a wrong assumption.
[2] The data has been extracted from Bostwick *et al.* [25].

3. GRAPHENE ON RUTHENIUM

there is no reason to believe, why it should vary so strongly in both cases. Indeed such a strong variation of the C-C bond is very unlikely.

- The diffraction of electrons from the buffer layer cannot be any indicator for the interaction of the graphene π_z orbitals with the buffer layer since the diffraction can only happen to the already emitted photoelectrons. So the electrons must be excited to vacuum level, if diffraction shall happen. If the buffer layer on the substrate was moved 20Å away from the graphene layer, diffraction of the emitted electrons would still be present, but there would be no interaction at all between the layers.

Therefore, the simplest explanation of the satellite bands is that the diffraction happens between the buffer layer and the substrate. This naturally explains, why the satellite spots show a hexagon that is rotated by 30°, since then the diffracted electrons arise from the sharp electron beam from the photoelectrons from the π-band, which are subsequently scattered on the lattice points[1]. Thus the orientation of the shadow spots equals the orientation of the lattice points of the reciprocal lattice.

It is important to understand that diffraction on the buffer layer would not reveal in such satellite spots, since the buffer layer has the same lattice constant as the upper graphene layer. Instead the spots can easily be explained be diffraction on the superstructure of the buffer-Ru(0001)-interface.

So how do the strong satellites go together with a weak interaction of the graphene layer with the buffer layer? In fact the strength of the diffraction satellites rather supports the theory of a weak interaction of these two layer, since the electrons can only be diffracted, if they are not caught in the crystal potential.

3.6 Summary

Our results show that intercalation of gold under graphene can be a useful technique to restore graphene's unique properties when bonding to a substrate strongly modifies its electronic properties. These changes can range from modest, such as changing the

---
[1] It is important to keep in mind that diffraction points refer to lattice points of the reciprocal lattice, which means that the M-point is situated in the middle of two diffraction points (see Fig. 1.3) and the Γ-M direction is 30° rotated with respect to the Γ-K direction.

3.6 Summary

natural doping of the graphene-substrate system, to more drastic, as in the case where the unique properties of the graphene are lost due to strong substrate hybridization as it is the case for the buffer layer on Ru(0001). Here no symmetry breaking can be observed, consistent with the fact that the buffer layer has the structure of graphene, but the strong bonding to the substrate destroys the conical bands [102]. This is a particularly important consideration for the interface between graphene and electronic contacts.

Moreover, the nature of the satellite bands that have been observed previously for graphene on SiC [25, 50] could be clarified. The present author agrees with the opinion in previous studies that they result from diffraction of the photo electrons, but they rather result from a diffraction between the buffer layer and the substrate and not between the graphene and the buffer.

In conclusion, the electronic structure modification of a single graphene layer on Ru(0001) upon the gold intercalation was studied by means of angle-resolved photoelectron spectroscopy. The spectral functions of the bilayer graphene film on Ru(0001) is characterized by the absence of any energy gap in the electronic structure. The energy gap for the π-states is found after intercalation of Au monolayer underneath monolayer graphene on Ru. The appearance of such gap in the electronic structure is assigned to the fact that the symmetry for two carbon sublattices is broken in the graphene/Au/Ru(0001) system due to the geometry of the system.

3. GRAPHENE ON RUTHENIUM

4

Graphene on SiC produced by Nickel Diffusion

4.1 Introduction

The graphitization of the SiC(0001) surface, by annealing at temperatures above 1150°C under UHV conditions, has been known for a long time [15] and has become one of the predominate methods for graphene fabrication in the field of experimental physics in the last ten years [23, 28, 46, 50, 55, 101, 109, 118, 119]. Although this high temperature graphitization method produces samples of such a good quality that e.g. quasiparticle-interactions in ARPES-data can be studied in detail [50], LEEM-measurements reveal highly non-homogeneous carbon coverage on these samples [51]. Recently discovered alternative methods include growth on the same surface *ex situ* under an argon atmosphere [53, 54] and graphene growth on the SiC(000$\bar{1}$) crystal face, where the quality of the graphene layers is no better [56] but the properties of single-layer graphene are preserved, due to the different stacking of the layers [55, 119].

A little-studied method of graphene growth on SiC utilizes the chemical reaction of nickel with the substrate. This process has been at the focus of research on metal-SiC contacts [124, 125, 126], graphite intercalation compounds (GICs) [120], nanostructures [122], and only recently has it been recognized in the field of graphene research [121, 127]. As illustrated in Fig. 4.1, graphene is not *necessarily* synthesized during the process and it has been reported that slight modifications of the reaction process can lead to e.g. different types of islands on the same sample [120, 128], graphene islands of

4. GRAPHENE ON SIC PRODUCED BY NICKEL DIFFUSION

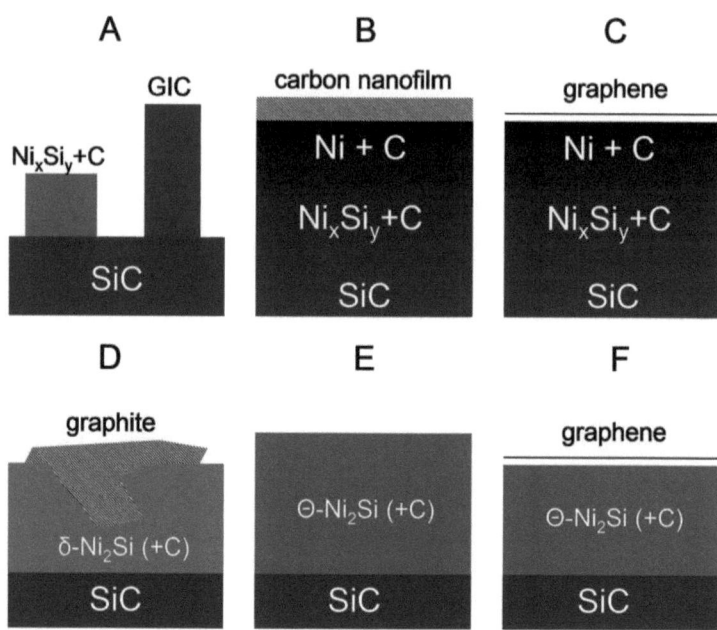

Figure 4.1: **Different configurations of the Ni-SiC-interface after annealing.** - Previous studies resulted in systems with remarkably different characteristics under only slightly modified preparation procedures. (A) A sample as investigated by Robbie et al.. After depositing 2.5 ML of nickel on the SiC(0001) surface and annealing at 600°C under UHV conditions, they found two different types of islands on the sample; one was considered to be a GIC [120]. (B) and (C) Samples with a pre-deposited 200nm Ni-film after a short annealing at 750°C and subsequent cooling with different cooling rates under vacuum conditions, according to Juang et al.. Depending on the cooling rate, they observed either a graphene layer or a disordered carbon nanofilm on top of their samples [121]. (D) Hähnel et al. pressed a Ni disc onto the 6H-SiC wafer under an argon atmosphere and annealed at 1245°C, which resulted in graphite clusters on a δ-Ni$_2$Si crystal. The graphite planes grew quasi perpendicular to the SiC(0001) plane [122]. (E) The study of Fujimura and Tanaka focusses on the reaction of 0.5μm nickel films on polycrystalline α-SiC (6H). Depending on the annealing temperature they obtained either δ- or Θ-Ni$_2$Si on top (Θ-type only above annealing temperatures of 1400°C) [123]. (F) This is the sample configuration as we produced it. See also Fig. 4.2.

mm size, disordered thin carbon-layers (both [121]), or bulk-internal graphite formation with the graphite planes quasi perpendicular to the SiC(0001) plane [122].

One of the most interesting features of this method is that all studies previously mentioned agree that the quality of the SiC surface does not seem to be too important for the quality of the resulting graphene layers or other carbon structures formed. Since this automatically leads to a cheaper production of the substrate (since the waver loss rate is automatically lowered), such an approach could be of special interest for future industrial production applications.

In this chapter, we use core-level photoemission and ARPES to study graphene produced by nickel diffusion on SiC. LEED, ARPES and core level measurements prove that the graphene layer is situated on a Θ-Ni_2Si substrate that got synthesized upon the nickel diffusion. The spectra reveal a strong hybridization of the graphene π-bands with the nickel d-bands comparable to the situation of graphene on nickel (see next chapter). Moreover, the core level and valence band spectra show the appearance of second-layer islands with strongly varying orientations. Our analysis of these spectra show that the bilayer islands prefer certain orientations that can be attributed to different superstructures resulting from the Moiré pattern of the rotated second layer islands.

4.2 Apparatus and Preparation

All measurements in this chapter have been performed at the BESSY end station described in Chapter 2 in the respective subsection. The end station was attached to the UE56-1 SGM beam line at BESSY.

The sample preparation was performed in the analysis chamber, where the base pressure never exceeded 5×10^{-10} mbar. We started with a 6H-SiC crystal, from SiCrystal^{TM}, with a hydrogen-etched (0001) surface, which was annealed at 600°C for several hours until sharp 1×1 LEED spots could be observed (see also Fig. 4.2). In some cases a $\sqrt{3}$ structure could be observed on the SiC surface before the Ni deposition, but no difference in the quality of the graphene layers deposited later could be detected.

The Ni deposition was performed via molecular beam epitaxy (MBE) from an electron beam nickel evaporator. Before every deposition cycle, the deposition rate was

4. GRAPHENE ON SIC PRODUCED BY NICKEL DIFFUSION

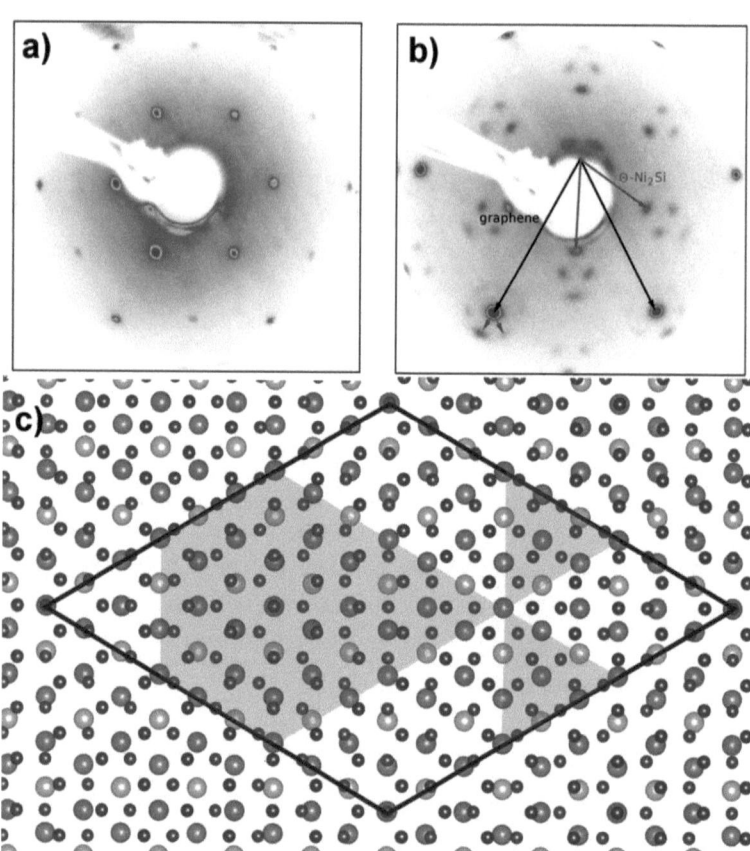

Figure 4.2: **LEED images and a model of graphene on Θ-Ni$_2$Si** - a) LEED image of the clean SiC 1 × 1 reconstruction before the Ni deposition, taken at 165 eV. b) LEED image of the graphenized sample, taken at 49 eV. The long black vectors belong to the graphene, the shorter grey vectors belong to the unit cell of the underlying Θ-Ni$_2$Si(0001) plane and the short grey vectors at the top of the black graphene vector belong to the 3√3 superstructure. All other LEED spots can easily be reproduced by linear combinations of these vectors. c) Model of the system with the superstructure drawn in. The upper Ni atoms dark grey, the lower Ni atoms are light grey, silicon is blue. The two sublattices of the graphene layer have slightly varying grey tones to illustrate that the hybridization does not occur exclusively in one sublattice. The shaded area in the unit cell of the superstructure shows the area in which one sublattice is closer to the nickel atoms, while in the non-shaded area within the unit cell, the other sublattice is closer to the nickel atoms.

4.3 Results

determined with a quartz crystal oscillator. During the determination process, the sample was turned away from the evaporator and was turned into the molecular beam only after the deposition rate was well stabilized. Subsequently, 10 to 20Å were deposited on the sample before it was rotated to face away from the evaporator, which then was switched off. Afterwards, the sample was heated resistively by direct current. The annealing temperature was around 1000°C and was estimated visually by the color of the sample[1]. We always annealed the sample for 10min. Annealing at lower temperatures resulted in an ordering of the nickel layer or intercalation of the nickel into the crystal (probably building carbon-rich δ-Ni_2Si), without any sign of graphene π-bands in the subsequently registered ARPES scans; higher temperatures resulted in an evaporation of the nickel and graphitization of the sample via the high temperature process[2].

Finally, it is important to mention that the thickness of the deposited nickel layer (10 to 20Å) had no influence on the spectra of the graphenized system that will be discussed in the following sections. Instead, an uncontrolled variation of graphene coverage was observed, which could be traced back to the only poorly-controlled variable, namely the annealing temperature. This is well supported by the fact that differences in sample characteristics could be observed over one and the same sample, when it had experienced a slight temperature gradient.

Measurements were performed with six different samples with predeposited Ni-layers of 10, 15, and 20Å thickness.

4.3 Results

As can be seen in Fig. 4.2, the LEED spots of the graphenized system show a clear, bright hexagonal spot pattern that can obviously be attributed to the graphene layer. This is supported by the size of a graphite-like peak in the core levels (see also Fig. 4.3), as well as by the strength of the graphene-typical π-bands, which will be discussed later. Interestingly, both LEED images in Fig. 4.2 were taken with the same sample

---

[1] Although this method might seem rather imprecise, the experience of the present author and collaborators could guarantee an error below 50°C, since the quality of the resulting graphene layer, assembled by the often-used *classical* graphitization of SiC, is extraordinarily temperature-dependent and thus served as a perfect test of practice.

[2] This could easily be determined by the graphene band structure, which showed all characteristics of graphene on SiC.

4. GRAPHENE ON SIC PRODUCED BY NICKEL DIFFUSION

orientation, proving that the graphene lattice has the same orientation as the SiC(0001) surface and is not rotated by 30°, as is the case when the graphene is grown by simple annealing at temperatures above 1150°C, where the growth process is responsible for the rotation of the lattice [48]. As will be discussed later, the underlying structure is a Θ-Ni_2Si crystal, and since the least amount of nickel deposited on the sample was 5ML, the Ni_2Si layer can be estimated as at least 7ML thick, which means that no SiC spots contribute to the LEED pattern.

As also explained in the caption of Fig. 4.2, the additional rather bright spots that are marked by red vectors belong to the Θ-Ni_2Si (0001) plane underneath the graphene layer. The blue vectors point to the additional spots, belonging to the resulting $3\sqrt{3}$ superstructure, as marked in Fig. 4.2 c). The respective supercell contains 8×8 graphene unit cells. All spots in the LEED images that do not directly belong to the structures discussed can easily be reproduced by linear combinations of two or three of the vectors referring to those structures.

4.3.1 Core levels

The core-level overview spectrum in Fig. 4.3 reveals a strong oxygen peak for both the clean SiC(0001) surface and the Ni-covered sample, at 533 eV binding energy. For the 1×1 SiC(0001) surface, this has been observed before [47], and is thought to result from an oxygen layer on top of the surface, which is a relic of the *ex situ* preparation [52]. The relative strength of the oxygen peak to that of the surface peaks increases after the deposition, showing that the nickel probably intercalates underneath the oxygen layer formerly situated on the 6H-SiC surface. However, after the annealing procedure, the oxygen signal is completely gone and the graphenized system shows no signs of other components than nickel, silicon and carbon.

The C1s core levels in Fig. 4.3 from the clean surface show a slightly asymmetric peak, similar to the respective curve after the nickel deposition. Both spectra can be fitted with two Voigt functions at 283.41 and 283.55 eV binding energy, showing that the slight shift of the photoemission intensity maximum in the curves results from a change in the relative concentration of different carbon configurations. The slight shift suggests that the predominant peak for the clean spectrum results from structural changes near the surface, while the weaker peak results from carbon deeper within the bulk, since these core levels would not change after the deposition and - although

4.3 Results

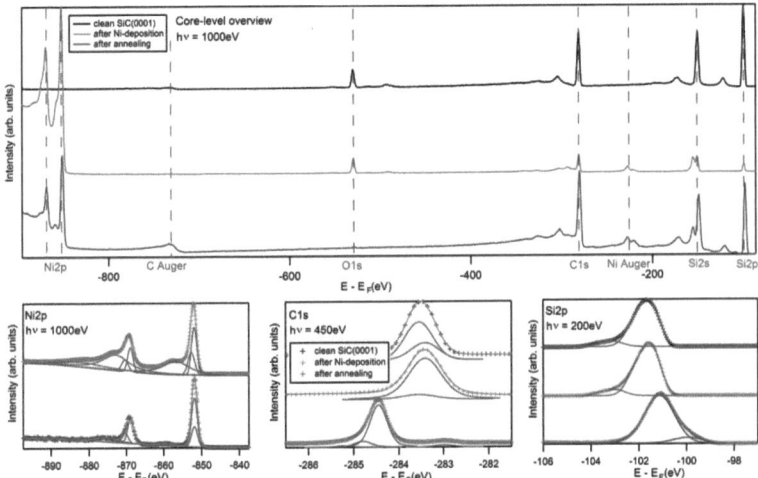

Figure 4.3: Core levels - Upper panel: overview of core level spectra taken at 1000 eV photon energy from the clean 1 × 1 SiC(0001) surface, as well as immediately after the deposition and after the annealing. The most important peaks are marked by the dashed lines. Due to the Auger- and core-level density at lower binding energies, not all peaks are marked in that range. Lower panels: Core level spectra of Ni2p, C1s and Si2p as marked, respectively. In the C1s- and Si2p-core level plots, light grey lines refer to the full fit functions, while the lines below represent the specific individual peaks. The C1s and Si2p core-level scans from the clean surface agree well with previous results [47, 48]. Although the fits of the Si2p spectra look as if the spin-orbit splitting was not taken into account, that is simply a result of the relatively small spin-orbit splitting in SiC. The fits for the Ni2p core levels are rather complicated, due to their satellites and the slightly different sites of the Ni atoms. Especially the spectra right after the deposition can be fitted only by assuming two different sites for the nickel, but the broad widths of the satellites make it difficult to resolve them in a reasonable manner.

4. GRAPHENE ON SIC PRODUCED BY NICKEL DIFFUSION

the overall C1s core level intensity was clearly weaker - if there were no structural changes, the relative height of both peaks should stay the same. This argument is also supported by the fact that the peak which was weaker after the deposition shifted to higher binding energies.

Following annealing, the C1s core level spectrum changed drastically and became very similar to that of graphene on SiC with a slightly asymmetric peak at 284.5 eV and a small bulk-related peak at 283 eV (which is about 0.5 eV lower in the case of SiC [56]). The maximum shifts by over 1eV towards higher binding energies, and the signal at roughly 283.5eV binding energy vanishes totally, while instead a small peak at 283eV appears. Although the large peak is well known from graphene on SiC(0001) [56] and is thus easily understood, the small peak must be due to carbon in a significantly different position. However, owing to the fact that there must be a large amount of spare carbon from the SiC that probably sits as *defects* in the Ni_2Si layer, it is most likely that this produces the additional C1s peak.

The Si2p spectra taken at 200 eV photon energy in Fig. 4.3 agree well with previous studies for clean SiC [48] and show an asymmetric structure, which results from the spin-orbit splitting (with a splitting of roughly 600 meV, which is thus not directly visible due to the peak width) on the one hand, and from the silicon atoms that are deeper within the bulk and do not belong to the 1×1 reconstruction on the other hand. The mean free path at 100 eV electron energy, taken from the universal curve, is in the < 1 nm range, which results in the observed low intensity of the observed bulk states. Only a very weak shift of approximately 60 meV to lower binding energies was found for the states that correspond to the 1×1 reconstruction.

After the annealing, drastic changes occurred in the core-level spectra of Si2p, and no SiC-typical signal could be observed. The intensity maximum shifted to 101 eV, which is a typical value for nickel silicides [129], with a small peak at 100 eV from the bulk.

The nickel 2p core levels, as well as their satellites (also in Fig. 4.3), can reveal much information about the system [130, 131]. For the Ni2p core-level spectra immediately after the deposition, the Ni2p peaks could not be fitted properly with only one Voigt line, but instead two Voigt lines close together (probably also due to surface effects, as for the other core level spectra) were required. However, in the case of the Ni core levels, this results in rather complicated fit functions due to the Ni2p spin-orbit splitting combined with the satellites, which were fitted with only one Voigt function each, due to their widths. The fits still reveal information about the rough distance from the main peak to its satellite, which lies between 5 and 6 eV (the higher Ni core-level peak is situated at 851.5 eV, while its satellite appears at slightly more than 857 eV), and thus agrees sufficiently well with previous data for metallic Nickel [130].

After the annealing, a proper fit of the spectral function could only be obtained by assuming that two slightly different spin-orbit splittings are present[1], which is reasonable since the nickel atoms have two different sites in the Θ-Ni_2Si layer. The satellite is then separated from its main peak by 8 eV, which refers, according to Nesbitt *et al.* [131], to a semiconducting or only barely metallic system. This supports the hypothesis discussed later in the text that the substrate is Ni_2Si, which shows only very weak DOS at the Fermi level [129]. Notable is also the sharpness of the peaks, suggesting that all the nickel is in the same system and no island formation has occurred, as suggested by Robbie *et al.* [120].

4.3.2 Valence bands

In Fig. 4.4, 1 to 3, the valence band spectra for samples of different graphene coverage are shown. As mentioned before, no dependence of the nickel coverage and the intensity of the graphene π bands could be observed, but a strong sensitivity to the annealing temperature was noticed. Thus, since we always estimated the annealing temperature by eye, an uncontrollable variable entered into the measurements; however the strength of the π bands in comparison with the other bands can be considered as a good indicator of the actual graphene coverage, which means that Fig. 4.4, 1 to 3, show samples with increasing graphene coverages. The valence-band spectra indicate clear π bands, proving the graphitic character of the system, with the π-band intensity maximum at the Γ point at 8.4 eV binding energy, which is a reasonable value if the system is compared to its most similar well-studied systems, graphene on silicon carbide and graphene on nickel (see chapter 5). Further comments on the nature of these bands, as well as doping, will be given in section 4.4.

An estimate of the graphene coverage achieved is rather difficult, since the formation of the graphene layer occurs simultaneously with the substrate formation and therefore no measurements on a clean Ni_2Si surface were taken. Thus, since the graphene

---
[1]Chemical influence on the spin-orbit splitting is not unusual. See Balasubramanian [132].

4. GRAPHENE ON SIC PRODUCED BY NICKEL DIFFUSION

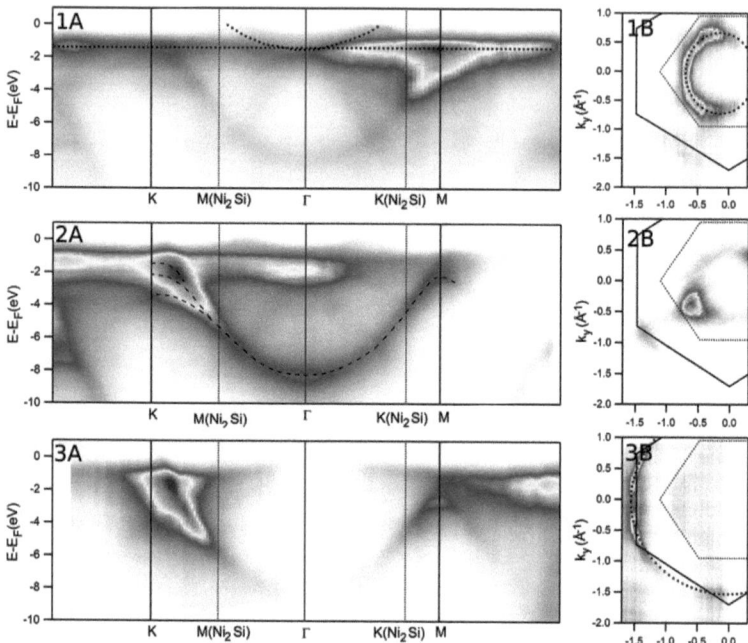

Figure 4.4: Valence band spectra - 1 to 3, A and B: The spectral function close to the Fermi level in Γ-K- and Γ-M directions and corresponding Fermi surfaces for samples with low (1), medium (2) and high (3) graphene coverage with BZs drawn in (more information in the text). The reason for the strong, and slightly confusing, intensity fluctuations in the Fermi surface plots lies in the generally low intensity at the Fermi surface, which makes the spectra more sensitive to camera- or channel-plate-induced intensity fluctuations. The weakness of the π-bands at the Γ-point in Fig. 3A is due to a non-optimal angular position during the scan.

coverage estimation process is based on some assumptions and is not straightforward, it will be the focus of subsection 4.4.2, where the results are discussed further.

However, a closer look reveals an extremely unusual spectral function of the graphene interface. The following extraordinary characteristics of the system can be noted:

- In Fig. 4.4, 1A and B, the features of the substrate are marked by blue lines and agree with previous ARPES studies on Ni_2Si [129].

- The graphene π-bands are strongly hybridized with the nickel d-bands, as shown by the dashed lines in Fig. 4.4, 2A. This situation seems to be similar to graphene on Ni(111) [42, 67] (see also next chapter).

- With high graphene coverage a non-hybridized graphene π-band passes through the Fermi level with a circular shape (see Fig. 4.4, 3B), unaffected by the boundary of the first BZ. As will be explained in the next section, this feature can be interpreted as the photoemission signal of varyingly oriented bilayer island atop the first graphene layer.

- The previously mentioned circularly shaped photoemission signal shows intensity variations as can be seen in Fig. 4.4, 3B. A particularly noticeable feature is the intensity minimum in Γ-M-direction.

- This π-band crossing the Fermi level shows a strong intensity loss before reaching it[1].

4.4 Discussion

4.4.1 The Graphene Substrate

To understand the electronic structure of the system, it is important to determine the actual substrate to which the graphene is bonded. Although it has been mentioned before that the substrate is Θ-Ni_2Si, the actual determination process has not yet been discussed.

The appearance of the nickel-derived d states at 1.6 eV binding energy suggests that the underlying substrate is Ni_2Si, since $NiSi$, $NiSi_2$ and Ni_3Si would have their

---

[1]The intensity loss occurs clearly *before* the band reaches the Fermi level and is not *due to* an incorrect estimation of the Fermi edge, which was determined from an integrated spectrum.

4. GRAPHENE ON SIC PRODUCED BY NICKEL DIFFUSION

d bands at significantly different binding energies (see tabulation 4.1). Although it is probable that carbon atoms are placed within the Ni$_2$Si-lattice (see also [122]), a nickel carbide as the substrate can be excluded for several reasons: Primarily, such nickel carbide products have not been found in any comparable study and actually would be in strong conflict with results concerning the phase diagram of this reaction [124, 126, 133]. Furthermore, a stronger substrate-related peak in the core level spectra of C1s would then be expected (see Fig. 4.3).

	Franciosi et al. [129]	Bisi et al.[1] [134]	Bylander et al. [135]
NiSi$_2$	3.15 eV	-	-
NiSi	1.8 and 3 eV	2 and 3 eV	-
Ni$_2$Si	1.3 eV	1.5 eV	-
Ni$_3$Si	-	-	0.3 eV

Table 4.1: **Binding energies of the Ni d bands in different nickel silicides**

Structures of the type Ni$_x$Si with $x > 3$ can be excluded, since their d bands would be at the Fermi level and such a system would not explain the resulting band structure, as will become clearer later in this section. Still, the Ni$_2$Si system can crystallize either in the orthorombic δ-Ni$_2$Si type or the hexagonal Θ-Ni$_2$Si type [136]. Experiments showed that the δ type usually only transforms to the Θ type at temperatures above 1200°C [136], but according to Tomam this temperature can be lowered to 800°C by the presence of silicon in solid solution [137]. Although Fujimura and Tanaka claim that the Θ phase in the present system can be realized only at temperatures above 1400°C [123] (which stands in fact in strong contradiction to the well-studied graphitization process that occurs at these temperatures and which has been extensively studied in the last few years [23, 49, 50]), we have strong evidence that the Θ phase is present in our case, as will be explained below.

In previous studies, where the Ni$_2$Si was synthesized in the orthorombic δ system, Hähnel et al. produced graphite planes that grew quasi perpendicular to the surface [122]. Since we did not observe such structures and the LEED images do not show any signs of anisotropic compounds (the δ type is strongly anisotropic, which necessarily would be visible in the LEED images and probably even in the ARPES-data), we can assume that the graphene is formed on a hexagonal Θ-Ni$_2$Si substrate. This conclusion is supported by the lattice parameter in the hexagonal (0001) plane, which amounts to

4.4 Discussion

3.805 Å [137] and is thus in good agreement with the observed lattice mismatch in the LEED images.

All together, this gives stronger arguments for the crystal structure shown in Fig. 4.2, which results in a superstructure of 8×8 graphene cells lying on a $3\sqrt{3} \times 3\sqrt{3}$ supercell. The mismatch, calculated from free-standing graphene and the bulk-truncated Ni_2Si, is in that case less than 0.5%. As can be easily seen, in this configuration neither of the two sublattices is more strongly perturbed, but instead both sublattices experience the same interactions with the substrate.

4.4.1.1 Θ-Ni_2Si as a substrate for graphene

Since the δ modification of Ni_2Si predominates in most systems, the overwhelming part of the literature is focussed on this configuration, and there barely exists literature on the Θ modification. However, the theoretical study of Peterson et al. predicts weak Si-p states crossing the Fermi level [138], which makes the material a metal with a very low DOS at the Fermi level. Our photoemission data show these states, in form of a parabola marked by the upper blue dashed line in Fig 4.4, subfigure 1A, and by the blue dashed line in subfigure 1B. This silicon state does not cross the graphene π-bands as far as one can see from our data.

Based on our knowledge of the system, we would expect the hybridization of the graphene π bands with the Ni d bands, as observed and indicated in Fig. 4.4, 2A. Since there is no reason to believe that not all the graphene that is directly connected to the substrate should show this hybridization, the fact that the π-bands, which cross the Fermi level (as shown in Fig. 4.4, 3B) do not exhibit any hybridization with the nickel d-bands, strongly suggests that this band reveals a second layer.

4.4.1.2 Structure and orientation of the graphene bilayer islands

The fact that the bilayer π-band crosses the Fermi level with a circular shape, partly outside the first BZ (see red dashed circle in Fig. 4.4, 3B), can be explained by a variety of different orientations for the bilayer islands. Such a configuration also naturally explains the photoemission intensity loss of these bands approaching the Fermi level, but it seems surprising that no such effect could be observed in the LEED images. However, if the bilayer coverage is low, which is consistent with the C1s core level spectra and the ARPES data, the LEED signal of these islands must be very low. If

4. GRAPHENE ON SIC PRODUCED BY NICKEL DIFFUSION

we interpret the small shoulder of the graphitic C1s core level peak in Fig. 4.3 as a signal from the second layer, the coverage would be below 0.1ML. Since LEED does not add the signal of all orientations but should show every orientation separated, it seems clear, that no signal can be observed with this method.

It is well known that graphite can stack in the most common AB-stacking mode [139], but also the turbostratic mode [139] or in the special case of graphite grown by silicon evaporation of SiC on the $(000\bar{1})$ face in a 27.8° rotated manner [55] (See Fig. 4.5). Commensuration of two graphene layers is possible in several stackings, but the exact binding energies for different stackings are difficult to calculate, due to the weakness of the interaction between the layers [140].

In Fig. 4.5 six commensurate bilayer graphene stacking modes are shown. Among the infinite number of possible stackings, these six modes are the ones with the smallest supercells. As will become clear below, in our case stackings with larger supercells are present. Each stacking mode can be characterized by the angle of rotation of the graphene unit cell of the upper layer compared to that of the lower layer. Therefore, this angle will be subsequently called the *rotation angle*. In the common AB stacking of the layers, the rotation angle amounts to zero. In the turbostratic stacking mode, the layers are rotated by 21.8°. The eight bilayer stacking configurations with the smallest unit cells are as follows

Graphene unit cells by supercell	Rotation angle	Commons
1	0°	AB or AA stacking
7	21.8°	turbostratic
13	27.8°	as grown on 6H-SiC$(000\bar{1})$ [55]
18	13.2°	-
31	17.9°	-
37	9.4°	-
49	16.4°	-
51	7.4°	-

Table 4.2: **Different possible bilayer stackings**

In Fig. 4.6 the photoemission intensity at 1.7 Å^{-1} distance from the Γ-point for different angular positions at the Fermi level is shown. 0° refers to the Γ-K-direction of the first layer. Since the first layer does not give any photoemission signal at the

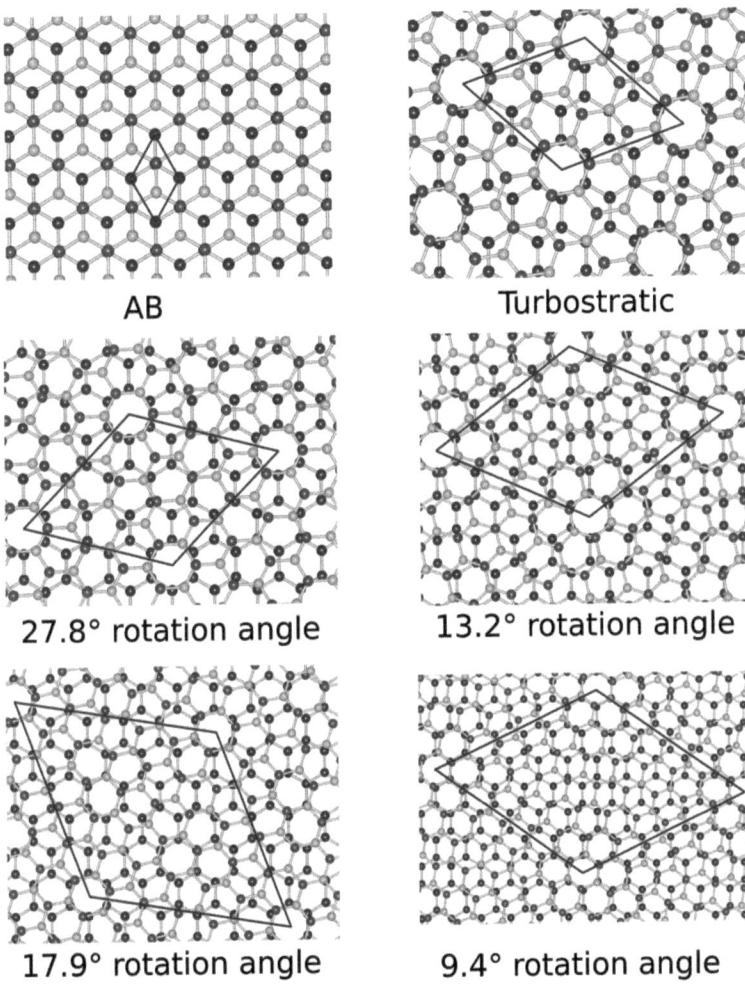

Figure 4.5: Different bilayer stackings. - The six different stackings indicated belong to the six smallest supercell configurations. The carbon atoms of the lower graphene layer are sublattice-dependent colored in different light grey-tones, the ones of the upper layer are in darker grey tones.

4. GRAPHENE ON SIC PRODUCED BY NICKEL DIFFUSION

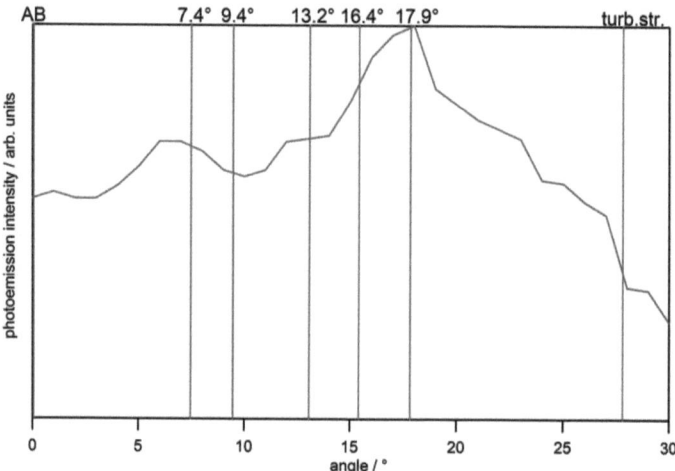

Figure 4.6: **Photoemission intensity of the bilayer band at** $1.7\ \text{Å}^{-1}$ **distance from Γ-point in angular dependence.** - Some rotational stacking positions for small unit cells are marked by lines.

4.4 Discussion

Fermi surface (see Fig. 4.4, 1B), the signal can be identified as coming from the bilayer. Since the second layer should only reveal electronic states at the Fermi level near the K-point, the strength of the signal can be seen as a direct indicator for bilayer islands with the respective rotation angle.

Although the rotation angle for the bilayer could vary between 0 and 60°, a plot from 0 to 30° is sufficient, since our data show a strict symmetry around the 30° rotation angle. This proves that the islands do not show any preferred chirality.

As one would expect, the photoemission intensity in Fig. 4.6 is lowest for 30° rotation angle, since the supercell would be infinitely large for this stacking[1]. The second obvious feature is an intensity peak around 17.9° rotation angle.

However, other stackings could not be directly resolved in our data. This is not surprising, since three factors are expected to limit the resolution for such a system:

- As previously mentioned, the van-der-Waal bonding between the graphene sheets is explicitly weak, which makes it difficult to calculate for different stackings [140]. Moreover, the system is very similar to graphene on nickel and one would expect that similarly to that system, the formerly unoccupied p_z-orbitals that are responsible for the van-der-Waals bonding lower the binding energy of the sheets, due to the charge transfer to the substrate [141]. This means that the system, depending on the island size, could form even larger supercells than the ones shown in tab. 4.2.

- The fact that many small islands of different orientations exist strongly suggests that the bilayer is of poor quality. This can lead to quantum confinement and other effects that lead to an unusual broadening of the bands [60].

- Finally, one would expect different distances of the islands to the first graphene layer for different orientations [140]. This automatically leads to varying doping strength, which will also give a broadening of the photoemission signal in Fig. 4.6, since the respective spectrum was taken at the Fermi surface and depending on the doping, the layers will give a slightly different signal at this energy.

---

[1] It has to be mentioned that small changes of the lattice parameter in the second layer can lead to the possibility of islands with this stacking.

4. GRAPHENE ON SIC PRODUCED BY NICKEL DIFFUSION

One additional remarkable feature in the data presented in Fig. 4.6 is the distinct peak at roughly 18°. It is suggestive to attribute this peak to the stacking with the 31-graphene unit cells superstructure that has a 17.9° rotation angle. However, our data and the lack of theoretical calculations make it difficult to explain such an unusual stacking preference, but it is probable that the growth mechanism itself is responsible for the stacking arrangement, since configurations with relatively small supercells (like turbostratic and 27.8° rotation angle – the latter is not drawn in in Fig. 4.6) barely exist.

Our data agree well with the recent work by Woodworth *et al.* [127], who achieved multilayer platelets atop the first graphene layer using a similar preparation method.

4.4.2 Layer Thickness

In Fig. 4.4, 1 to 3, clearly different layer thicknesses were obtained, as can be readily determined by the different apparent intensities of the π bands. The sample in Fig. 4.4, 1, shows a strong inner circle, which can be attributed to the substrate, and only weak π bands with no bilayer-induced intensity, which can only be the case if the graphene coverage is less than 1ML. The strength of the inner circle may suggests a coverage of about 0.7 ML.

The weakness of the inner ring in Fig. 4.4, 2, and the appearance of bilayer-induced intensity suggests that the coverage achieved is about 0.9ML, with islands of lower and higher coverage, probably comparable to the situation of graphene grown *in situ* by the high-temperature method on SiC [51].

The vanishing inner ring in Fig. 4.4, 3, suggests that the coverage exceeds 1 ML all over the sample, and can be roughly estimated as 1.1 ML.

In principle the similarity to the graphene/Ni(111)-system could lead to the conclusion that a closed bilayer growth is difficult to achieve. In fact, since the carbon from the graphitization results from the bulk could be the main reason that bilayer growth could be performed in this system. Since graphene on Ni(111) is usually grown *via* chemical vapor deposition, no closed second layer can be built, since the single carbon atoms do not stick to the surface.

4.5 Conclusion

In summary, we have successfully created a monolayer of graphene on SiC via intercalation of Ni atoms which formed Θ-Ni_2Si and allowed the carbon to segregate to the surface. Depending on the annealing temperature, bilayer islands are formed atop the first graphene layer. The islands show strongly varying orientations, which can be correlated with a lowered local density of states of the p_z orbitals perpendicular to the graphene sheet.

To our knowledge we are also the first to have produced a graphene layer on a Θ-Ni_2Si (0001) surface, and we are likewise the first to find a system where the graphene appears in such varying orientations.

Local probe experiments might lead to novel discoveries.

4. GRAPHENE ON SIC PRODUCED BY NICKEL DIFFUSION

5

Graphene on Nickel

5.1 Graphene on nickel

5.1.1 A short introduction to graphene on nickel

The formation of graphene layers on Ni(111) surfaces via chemical vapor deposition (CVD), by the cracking of hydrocarbons, has long been a known method [18, 68] and has drawn special attention because of the small lattice mismatch between free-standing graphene and the planar interatomic distance of the close-packed Ni(111) surface, which is only 1.3% [142]. This small lattice mismatch allows the graphene to form huge layers without defects. Furthermore the strong interaction of the graphene layer with the substrate limits the graphene thickness to one ML [67], making this fabrication method very promising for industrial purposes[1].

However, since the electronic properties of graphene on nickel are strongly modified by the hybridization of the graphene π-bands with the nickel d-bands [64, 65, 67], for most electronic applications the graphene layer has to be detached from the substrate. Recently this has been done with ultra-large graphene sheets of cm^2-size, made by CVD on nickel [37]. These sheets were prepared by CVD on thin nickel films on silicon wavers and then detached by etching the substrate.

---

[1]The self-inhibiting growth process is thought to result from the low electron density atop the graphene layer on nickel, since the formally empty π^* orbitals are occupied and participate in the charge transfer to the substrate [141]. However, Odahara et al. saw a LEED signal from a graphene bilayer system that had been detached by etching from a Ni(111)-surface. The authors suggest that the bilayer was formed accidently by crumbling of the detached graphene layer [143].

5. GRAPHENE ON NICKEL

Still, the modification of the graphene band structure arising from hybridization with the nickel 3d-bands is of high interest. ARPES serves well for studying the interaction of graphene with its substrates, but the hybridization of the Ni-d-bands with the graphene π-bands could as yet be observed clearly only for the lower graphene π-bands [65, 67]. Theoretical studies predict hybridized nickel states close to the Fermi-level at the K-point [42], and also at the M-point, due to hybridization of the graphene π-states with the nickel sp-bands (see also the theoretical band structure in Fig. 5.6) [42, 144].

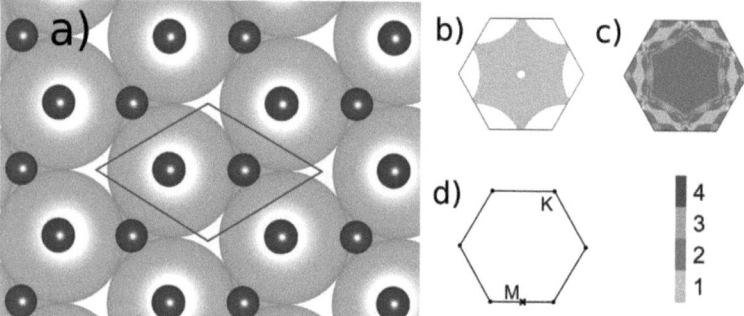

Figure 5.1: Graphene on Ni(111) - a) Model of graphene on Ni(111). The grey spheres atop represent the two carbon sublattices. The larger spheres underneath represent the nickel atoms. The unit cell is drawn in. b) & c) The Fermi surface for the majority and minority spin direction of the Ni(111) surface. d) Fermi surface of free-standing graphene. Subfigures b) to d) were taken from Karpan et al. [42].

The atomic structure of graphene on Ni(111) (as proposed by Gamo et al. from LEED-experiments [142], and corroborated by Kawanowa et al. with ion-scattering experiments [145]) is shown in Fig. 5.1. As one can see, the carbon atom of one sublattice is always situated over a hole, while the carbon atom of the other sublattice is situated over a nickel atom. This configuration is usually referred to as AC, while the later mentioned graphene on Ni(111) configuration, where the graphene sublattice symmetry is not broken, is usually referred to as BC (A then means the position right above a nickel atom, B means the position above a nickel atom of the second layer and C above a nickel atom of the third layer). The resulting violation of the sublattice symmetry in the AC configuration leads to a gap opening, which is difficult to observe since the hybridization dominates the electronic structure of the system[1] [65].

5.1.2 Graphene as a spin filter

Another interesting feature of multilayer graphene or graphite on Ni(111) is its theoretically predicted ability to serve as a perfect spin filter [42, 146]. The observation of giant magnetoresistance (GMR) in interlayer systems of ferromagnetic (FM) and non-magnetic metals (NM) [147, 148] has given rise to new applications, such as storage devices [149], as well as to the relatively new field of spintronics. A perfect spin filter would allow all electrons of one spin direction to pass through the device while the other spin direction is completely blocked, thus resulting in a perfectly filtering GMR-device with maximum conductivity. Not only FM|NM|FM systems can be used as spin filters, but the non-magnetic metal can also be replaced by an insulator (I) or semiconductor (SC) [150]. Even though spin filters are already utilized in hard disc reader heads, the state-of-the-art devices still function as a far from perfect spin filter, due primarily to material restrictions [150, 151, 152].

The reason why graphene and graphite might function as perfect spin filters becomes clear by looking at the spin-dependent DOS on the Ni(111) Fermi surface (see fig. 5.1) as calculated by Karpan *et al.* [42]. The majority spin direction in nickel has no states at, or near, the K-point, which is the only point in reciprocal space where graphene has a non-vanishing DOS at the Fermi level. However, due to the hybridization, spin-mixed states appear in the band structure, which strongly reduces the spin filtering for one ML of graphene on Ni(111). As one would expect, this effect decreases with more graphene layers. Karpan *et al.* calculated that nearly perfect spin filtering should be possible starting from a thickness of four graphene monolayers atop the nickel surface [42]. Although such a system might be difficult to be experimentally realized since the CVD-process on Ni(111) is self-inhibiting to one monolayer[2], further fine-tuning of the

---

[1] One might think that a gap-related discussion fails in the case of strong hybridization near the K-point. This is wrong in the case of graphene, since a non-broken symmetry of the two sublattices will still lead to a Dirac-crossing as theoretically calculated by Karpan *et al.* for the so-called BC-configuration of a graphene layer on Ni(111) [42] (see also Fig. 5.6).

[2] In contrast, the formation of multilayer graphene on polycrystalline nickel surfaces has been successfully performed, but only graphene flakes of comparatively small sizes and varying thickness could be obtained [153, 154].

5. GRAPHENE ON NICKEL

spin-filtering system could be done by the intercalation of other metals, which has been previously performed and can easily be realized experimentally[1] [110, 111, 155, 156].

In this chapter high quality ARPES data of graphene on Ni(111) are presented. A nickel layer has been deposited on these samples to check if spin filtering could be observed in our ARPES data, but no evidence for such an effect could be found. Furthermore our data sets show hybridization features of the upper graphene π-band close to the Fermi level that stand in excellent agreement with theoretical studies [42, 144].

5.2 Apparatus and Preparation

All measurements have been performed with the BESSY end station as described in subsection 2.2.2. The end station was attached to the UE56-1 SGM where the majority of the data were taken. Additional measurements were performed at the Max-Planck beam line UE56-2. All data sets shown in this chapter were taken with linearly polarized light to prevent spin-dependent effects in the measurements.

To achieve a high quality Ni(111) surface, 200Å of nickel were deposited by CVD on a W(110) surface that had been cleaned by several cycles of annealing at 800°C under an oxygen atmosphere of 10^{-7}mbar for several hours, followed by subsequent flashing at about 2200°C by electron bombardment. The growth of nickel on tungsten surfaces has been studied for decades [157], and W(110) is known as an ideal substrate for the growth of epitaxial layers since the high surface energy and the closed bcc (110) structure allow a high mobility of the condensing atoms, while chemical reactions between the substrate and the adsorbate are inhibited[158, 159, 160]. The nickel film was chosen to be relatively thick with 100 ML to prevent surface defects, induced by the lattice mismatch between the W(110) and the Ni(111) plane. After deposition, the crystal was slightly annealed at approximately 400°C for two minutes to order the nickel film and create a quasi defect-free Ni(111) surface.

To grow the graphene layer atop, a propylene atmosphere of 10^{-6}mbar was introduced in the chamber and the crystal was annealed at 600°C to induce cracking. Afterwards the chamber was pumped down to base pressure and the crystal was cooled.

---
[1]see also chapter 3.

For the formation of the nickel/graphene/Ni(111) interlayer system, one more Ni monolayer was deposited on the sample.

5.3 Results and Discussion

5.3.1 Band maps and energy distribution curves

In Fig. 5.2, the photoemission intensity maps of the systems measured are illustrated. The flatly dispersing Ni-d-bands of the clean Ni(111) surface are clearly visible at both photon energies used for the scans. Clearly the nickel d-bands are better resolved in the scans with the graphene layer atop the surface, although one might expect the opposite, since scattering with the photoelectrons coming from the nickel should be avoided. However, the high reactivity of nickel gives the clean surface a short lifetime, while the graphene layer covers and protects the surface[1] [156].

In the band maps as well as in the energy distribution curves (EDCs) in Fig. 5.3, one can see that the flatly dispersing nickel d-bands have their major intensity maxima between the Fermi level and 3eV binding energy. The intensity of the extremely flatly dispersed nickel surface state, as first reported by Himpsel et al.[161], which is best visible around the Γ-point and is marked by the black arrow in the upper left panel in Fig. 5.3, proves the cleanliness of the Ni(111) surface. The surface state could not be resolved in the scans taken at 100 eV photon energy. The accompanied broadness of the Ni-d-states suggests that the surface was dirtier during this scan.

The dark red ellipse in the upper right panel of Fig. 5.2 marks a nickel d-band that seems to partly vanish after the graphene layer has been deposited. Such a behavior can easily be explained by a hybridization of the respective nickel d-band with the upper graphene π-band. The proposed hybridization could not directly be detected in our data, since it takes place above the Fermi level. However, such hybridized nickel d-states are predicted by theoretical DFT calculations [144, 42] and are in good agreement with the EDCs (Fig. 5.3) and constant energy maps presented in Fig. 5.8.

The dashed ellipsoid in the upper left panel in Fig. 5.3 marks the non-dispersive 6eV satellite signature of the Ni d-bands, which was first observed by Hüfner and Wertheim

---

[1] Although one high quality scan takes usually approximately one hour, the quality of the surface decreased most during the sample transfer from the heating stage to the goniometer. This was verified several times, since the first photoemission intensity maps were always accumulated while the sample was still in the heating stage.

5. GRAPHENE ON NICKEL

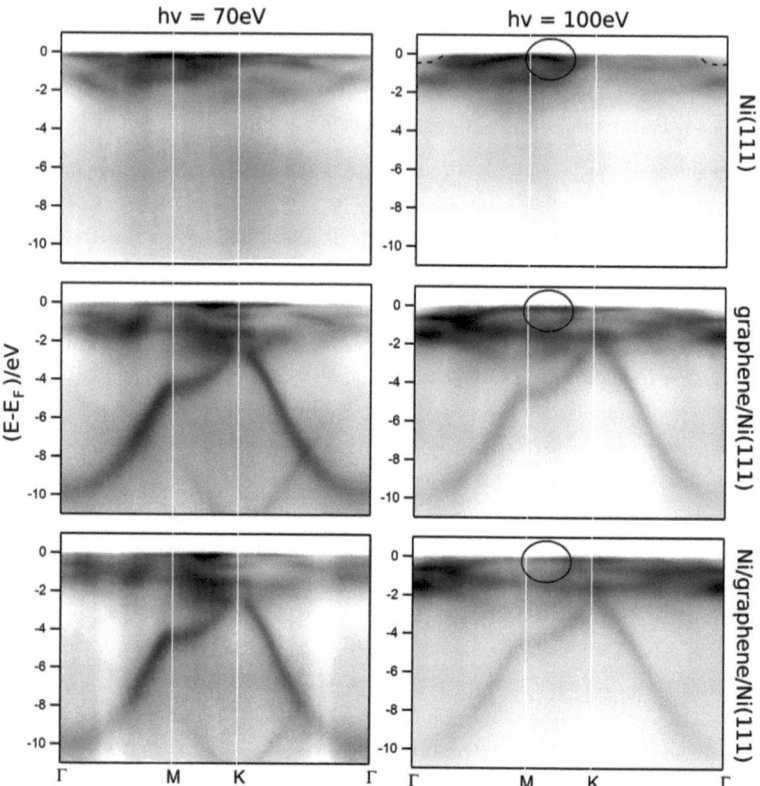

Figure 5.2: Spectral functions of Ni(111), graphene on Ni(111) and the Ni/graphene/Ni(111) interlayer system. - Left column: Band maps taken at 70eV photon energy along Γ-M-, M-K and K-M directions of the clean Ni(111) surface (upper row), one graphene ML on Ni(111) (mid row) and the Ni/graphene/Ni interlayer system (lower row). Right column: Band maps taken of the same samples respectively at 100 eV photon energy. The ellipse marks a nickel d-state (only visible in the upper panel), which becomes hybridized with the upper graphene π-band and thus moves above the Fermi level. More information in the text.

5.3 Results and Discussion

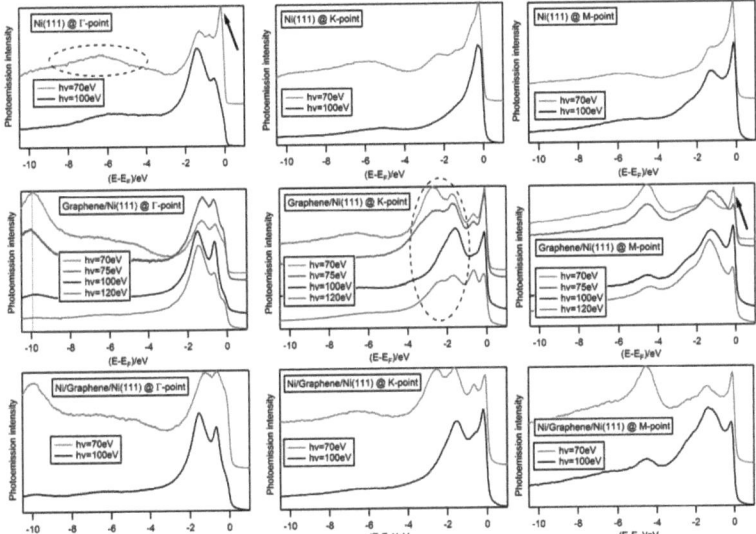

Figure 5.3: Energy distribution curves of the samples at the high symmetry points Γ, K and M. - Sample configurations, position in K-space, as well as photon energies are given in the insets. The ellipse in the upper left panel marks the non-dispersive nickel satellite at 6eV [162]. The same spectrum shows a strong nickel surface state (marked by the arrow). The middle row presents EDCs of the graphene on nickel(111) system with a strong π-state (marked by the blue dotted line in the left panel) and hybridized states (marked by the ellipse in the middle, and the arrow in the right panel). More information in the text.

5. GRAPHENE ON NICKEL

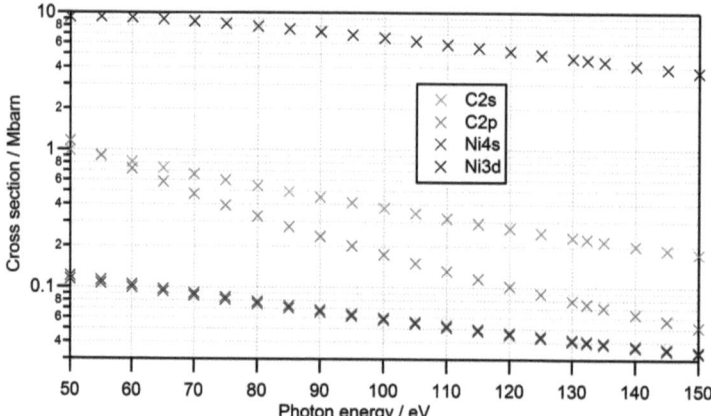

Figure 5.4: Cross sections for the nickel and carbon valence and nearby valence states. - The calculated cross sections of the atomic states that participate in the respective valence band spectra from http://ulisse.elettra.trieste.it/services/elements/WebElements.html. It has to be mentioned that calculations show that nickel $3d$-states of the pure Ni(111) surface are hybridized with the lower lying Ni $4s$ states, as well as with the Ni $4p$ states above the Fermi level, which leads to modifications of the cross sections [144].

5.3 Results and Discussion

[162] and explained by Guillot *et al.* [163]. The weakly dispersive nature that appears in this structure at 100eV photon energy can easily be traced back to the overlying 4s-bands [144]. However, as shown in Fig. 5.4 the cross section of the 4s-states is smaller by a factor of 100 in comparison to the Ni-3d-states. This results logically in a dominance of the nickel 3d satellites. The relative strength of these is best proven by the EDCs at the K- and M-point shown in Fig. 5.3, upper row, mid and right panel. According to Bertoni *et al.* the band width of the 4s-bands is strongly decreased at these points and the peak maximum is shifted by at least 2eV to lower binding energies [144]. Our data do not show such an effect at these high symmetry points, which proves that the respective feature cannot derive from the 4s-related states.

The formation of the graphene layer leads to dramatic changes in the band structure as can best be seen in Fig. 5.2, mid row, in the scan with a photon energy of 70 eV, where the characteristic graphene π-bands are clearly visible. As in previous studies of graphene on nickel, no Dirac cones could be observed. The electronic structure of graphene is strongly modified by the hybridization with the nickel 3d states around the K-point. The π-band reaches it minimum at a binding energy of 9.9 eV (as marked by the blue line in the respective panel in Fig. 5.3) at the Γ-point with a FWHM of 1.6 eV. Both results are in good agreement with previous experimental studies that usually show the photoemission intensity maximum from the π-states at slightly above 10 eV [65, 67]. The quality of our data sets is superior to previous ARPES studies of graphene on Ni(111) [67, 155, 156] - as proven by the rather well-resolved nickel d-states - and our π-band position stands in perfect agreement with theoretical calculations [42, 144]. After the deposition of one more nickel monolayer, the π-bands are slightly broadened with an FWHM of 1.9eV at the Γ-point, which is reasonable due to expected scattering of the photoelectrons in the upper nickel layer.

In Fig. 5.5 one can see EDCs at the high symmetry points in a different arrangement than in Fig. 5.3. As best visible in the EDCs at the K-point, the appearance of the graphene π-bands is accompanied by a shift of the Ni 3d satellites by approximately 700meV to higher binding energies (the maxima are marked by the dashed blue lines). This effect has been observed before [43] and is not in the scope of this thesis.

As one would expect, the EDCs in Fig. 5.5 for the Ni/graphene/Ni(111) interlayer system show decreased graphene π-band signals which is best visible at the Γ-point.

5. GRAPHENE ON NICKEL

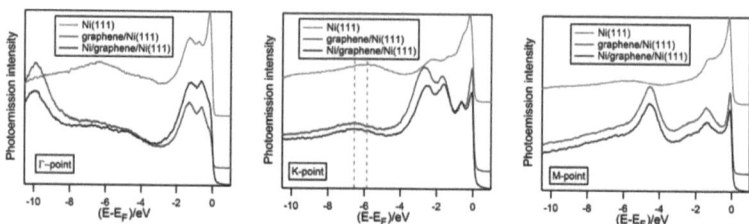

Figure 5.5: **EDCs at the high symmetry points for the measured systems taken with 70eV photon energy.** - The dashed lines mark the 500meV shift of the Ni $3d$ satellite peak.

Only the EDCs at the M-point show no clear reduction of the graphene-related photoemission signals, with the exception of the hybridized signal at the M-point close to the Fermi level, which will be discussed later in this chapter.

5.3.1.1 Spin filtering effects

Our measurements on the Ni/graphene/Ni(111) interlayer system could in principle serve to prove spin filtering by a lower photoemission intensity from the upper nickel layer than one would expect, if possible spin filtering was neglected. This would be feasible, since the upper nickel layer is grounded through the graphene layer, which is in turn grounded via the lower nickel film and thus the photoemission intensity of the upper nickel layer can serve as an indirect transport measurement. Since spin filtering should take place through the graphene layer, the photoemission intensity of the nickel d-bands of the upper layer should be slightly reduced (approximately 20% filtering of the majority spin direction was calculated by Karpan et al. [42]). However, the density of nickel d-band states makes it impossible to directly resolve the spin splitting in our data sets.

If we assume that minority and majority spin directions participate equally to the photoemission intensity, one would expect a reduction in the intensity of 10% for the upper layer[1]. This effect is decreased even further, since the mean free path is still high enough to allow the electrons from the lower nickel layers to contribute to the

---

[1] As mentioned this includes the assumption that minority and majority spin directions participate *equally* to the photo emission intensity, which is not the case directly at the Fermi level where the minority Ni d spin states lie partly above the Fermi level

photoemission spectra. Additionally, it seems likely that the nickel atop the graphene layer is only very weakly bound to the graphene[1], since all the binding electrons of the graphene layer are already occupied, and subsequently forms islands (this is in perfect agreement with previous studies [43, 141]), which could explain the reduction of sharpness in the photoemission signal. Consequently an evaluation of our data to find conclusive evidence for spin filtering effects remained unsuccessful.

5.3.1.2 Hybridization effects in detail

The slight modification of the EDCs, taken at 70 eV and 100 eV photon energy at the K-point for clean Ni(111), can be attributed to the change of position in the BZ in $k_\perp$ direction. With the graphene layer atop, the hybridization of the graphene π-states with the nickel d-states becomes clearly visible in the respective band maps in Fig. 5.2 and stands in good agreement with previous studies.

The EDCs in Fig. 5.3 serve even better to analyze the hybridization in detail. This is particularly the case since the ratio of cross sections of the nickel $3d$-bands to the graphene π-bands strongly increases towards higher photon energies within our photon energy range (from 12 at 70 eV photon energy to 20 at 120 eV photon energy). The EDCs at the K-point of the graphene/Ni(111) system show strong photoemission signals obviously induced by the graphene-layer (marked by the dashed ellipse in Fig. 5.3). Interestingly the lowest peak decreases strongly with higher photon energies, while the peak at -2eV binding energy still is very strong in the 100 eV photon energy EDC. This can only be explained by a π hybridization of the respective $3d$ state, while the peak below is a π-dominated peak.

Furthermore the peak closest to the Fermi level in the same panel shows decreasing relative photoemission intensity with increasing photon energy. This proves that it is a d-hybridized π-state. This state has not been observed before, but was predicted in theoretical studies[42, 144] and stands in good agreement with the data presented in the next subsection.

---
[1]This does not contradict the strong interaction between the graphene and the Ni(111) surface, since as the p_z orbitals are now hybridized with the underlying nickel d-states the probability of presence of the electrons on the graphene-covered surface is strongly reduced, which limits the charge-transfer from the graphene layer to the upper nickel flakes.

5. GRAPHENE ON NICKEL

Interestingly such a hybridization effect of the nickel d-state, which is closest to the Fermi-level also appears at the M-point (marked by an arrow in the right panel in mid row of Fig. 5.3 with a strong signal at 70 eV photon energy), although one would naively not expect such a hybridization at the M-point, since no graphene π-states are found here close to the Fermi-level. However, it has been theoretically predicted by Karpan et al. and Bertoni et al. that the graphene π-states hybridize with the M-point crossing nickel sp-states above the Fermi level [42, 144]. The calculations from Karpan et al. are shown in Fig. 5.6. The observed effect can be well explained within this model. The relative increase of this state at 100eV results from the additional hybridization from the nickel d-states as shown in the theoretical calculations for the AC-stacking in Fig. 5.6.

The hybridization is not easily observable in the EDCs of the Ni/graphene/Ni(111) system. This is obviously the case due to scattering processes in the upper nickel layer, but can also be attributed to the previously mentioned low binding energy from the surface to the upper nickel layer, which likely makes the nickel form islands on the surface. This has been observed with STM (M. Fonin, private communication)[1].

In Fig. 5.6, scans for the specific region of interest are shown. As will be shown below, agreement with the theoretical calculations of Karpan et al. exists, although the broadness of the bands resulting from the short hole life time make a detailed comparison impossible. Nevertheless, as expected our measurements show no appearance of Dirac cones, ruling out the BC configuration.

However, in accordance with the theoretical calculations we see several broad photoemission intensity maxima in the shown interval at the K-point:

- Directly at the Fermi level, a broad band is visible at all photon energies. Since from the theoretical calculations the nickel minority spin d-states, as well as the majority spin hybridized graphene π-band should contribute to the photoemission signal, it is not surprising that the band is relatively strong for all photon energies.

- According to Karpan et al., the majority spin states should give a maximum at 0.8 eV binding energy, while the minority spin states should give a maximum at

---
[1]Fonin tried to do an STM study of the Ni/graphene/Ni(111) system. The upper nickel atoms were so weakly bound to the graphene layer that the STM tip permanently moved the atoms, which tended to form islands.

5.3 Results and Discussion

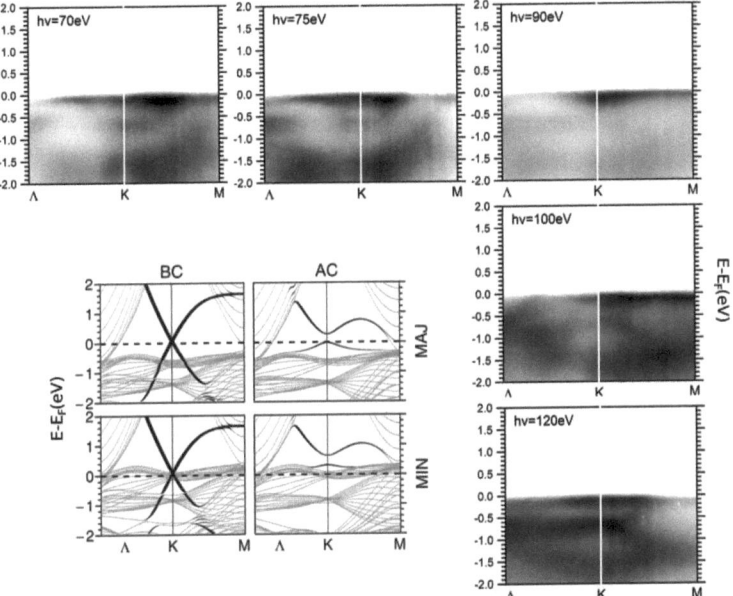

Figure 5.6: The spectral function around the K-point for graphene on Ni(111) - Upper and right panels: Bandmaps around the K-point taken at different photon energies. As can easily be seen at 90eV, the relative photoemission intensity of the hybridization of the upper π-band with the nickel d-bands is the strongest. Other two panels: calculated band structure for majority and minority spin direction from Karpan et al.[42]. More information in the text.

5. GRAPHENE ON NICKEL

0.9 eV. Although these peaks cannot directly be resolved in our data sets, the band in the ARPES data is slightly shifted to higher binding energies for lower photon energies. Additionally, the shape of this band at 70 eV photon energy is very similar to the theoretically predicted majority spin band, while the minority spin band shape is more similar to the one at 120eV photon energy. This can be seen as a direct indicator for the spin-dependent hybridization as predicted by theory, since one would expect that the cross section for the hybridized majority spin states moves to lower photon energies.

- The lower lying majority spin state at 3.2 eV binding energy is present at all photon energies.

The only scan in which the photoemission intensity at the Fermi level at the K-point clearly dominates the other bands is the one taken at 90 eV photon energy. This can be seen as a clear indicator that this hybridized π state has the highest relative cross section here, since the relative intensity of the other states is significantly reduced.

5.3.2 Fermi surfaces and constant energy maps

In Fig. 5.7 the Fermi surfaces of the measured systems are shown. These are the Fermi surfaces that are taken from the same scans as the band maps previously shown in Fig. 5.2. Nearly all Fermi maps show only broad features and reveal a poor k-space resolution in the measurement, which can easily be explained by the nickel induced high DOS close to the Fermi level. This leads to a broadening of the bands in terms of energy, resulting from the short lifetime of the holes. The broad band distribution indirectly enters the k-space resolution, since the bands are only weakly dispersive.

Still, two structures stand out in the scans taken at 100eV photon energy: one hexagonal structure, marked by the red line in Fig. 5.7, and a roughly circular structure closer to the Γ-point, marked by the dashed white line in the same figure. While the hexagonal structure looks similar to the theoretically predicted minority spin d-band structure, as shown in Fig. 5.1 c), the circular structure is neither shown in the theoretical minority nor the majority spin Fermi surface in this figure. In Fig. 5.2 the respective band is marked by the dashed curves close to Fermi level around the Γ-point in the upper right panel. Here the structure looks like the strongly-dispersing

5.3 Results and Discussion

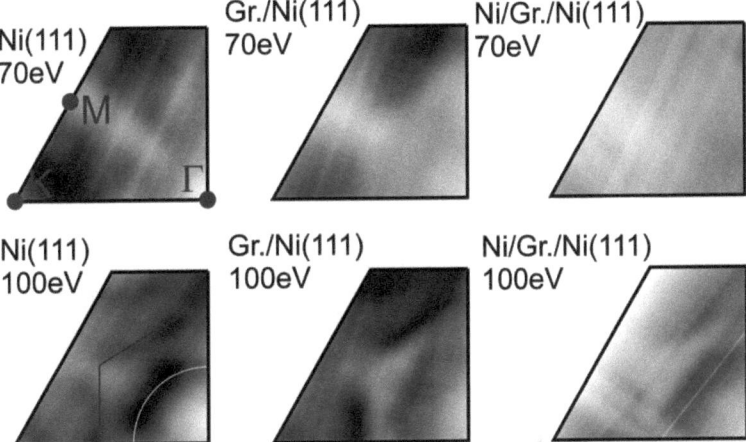

Figure 5.7: Fermi surfaces of Ni(111), graphene on Ni(111) and the Ni/graphene/Ni(111) interlayer system - A quarter of the first BZ on the Fermi surface of the different systems. High symmetry points are drawn into the upper left subfigure. The diagonal stripes are from the CCD camera.

5. GRAPHENE ON NICKEL

sp-bands from the majority spin direction, as predicted in the previously mentioned studies [42, 144].

In the 100 eV scan of the graphene on Ni(111) sample, fuzzy photoemission intensity maxima appear at the K-points. The facts that these maxima are well visible at 100 eV and only appear within the graphene system are proof that they must result from weakly π-hybridized nickel d-states, which stands again in perfect agreement both with previous theoretical studies [42, 144] and the data presented in Fig. 5.6.

As observed the highest relative cross section of the hybridized π-state at the K-point directly at the Fermi-level is located at 90 eV photon energy. The respective constant energy maps close to E_f are shown in Fig. 5.8. At roughly 200 meV binding energy the photoemission signal is most limited to the K-points, while at the Fermi surface roughly homogeneous photoemission intensity could be observed all along the K-K'-line. This agrees almost perfectly with the calculated band structure by Karpan et al. as shown in Fig. 5.6; since non-hybridized nickel d-states of the minority spin direction are located around the Fermi level, the π-state should be best resolved at roughly 200 meV below the Fermi level.

At 500 meV below the Fermi level no π-hybridized states can be observed and non-hybridized nickel d-states dominate the structure. These constant energy maps perfectly show that no BC configuration of the graphene layer on the Ni(111) surface can have occurred since this would lead to strong π-state signals even down to 1eV binding energy.

5.4 Summary

To conclude, we successfully grew a graphene layer on the Ni(111) surface and deposited one ML of nickel onto it to produce the nickel/graphene/Ni(111) interlayer system. As expected, no direct evidence for a spin filtering effect in this system could be found. However, our Fermi and constant energy maps of these systems show hybridization effects close to the Fermi level which have not been previously observed.

These effects are expected to be the case for a hybridized state right at the Fermi level at the M-point, as well as a strong hybridization at the K-point. The measurements presented here stand in almost perfect agreement with revious theoretical predictions [42, 144].

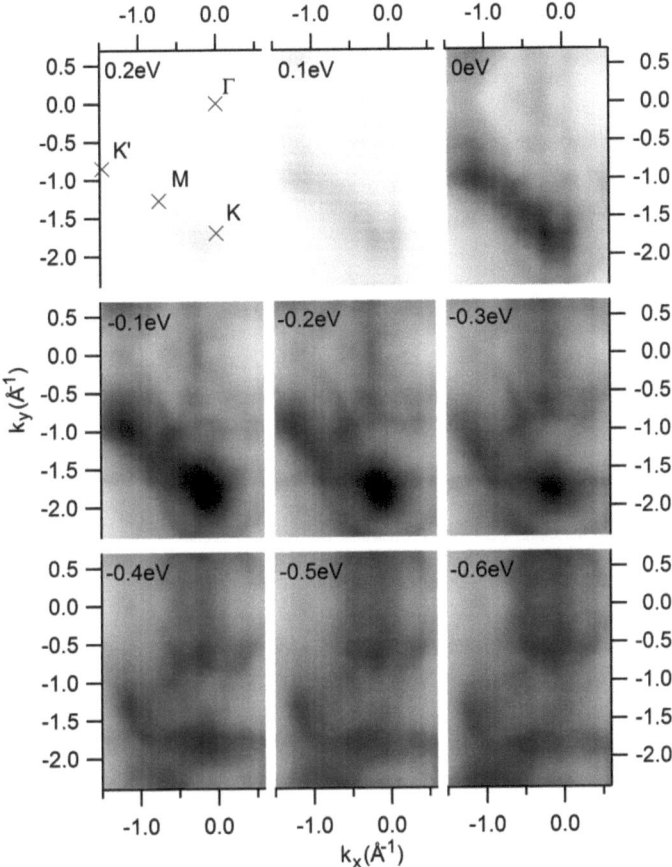

Figure 5.8: Constant energy maps of one monolayer of graphene on Ni(111) - The data was taken at a photon energy of 90 eV. Γ- and K-points are drawn in the upper left panel. Binding energies are inserted.

5. GRAPHENE ON NICKEL

Finally, our data confirm that the graphene/Ni(111) system is stacked in the so-called AC-manner.

6

Summary, Conclusions and Outlook

6.1 Summary

There is no doubt that graphene is considered the most promising modern material for a variety of future applications. In any graphene-based device, the graphene layers will be situated on substrates and electronic contacts to other materials are unavoidable. Therefore, studying graphene's interaction with different materials is an important step on the way to graphene-based electronics. Moreover, it is important to study different growth mechanisms of graphene layers, since different growth methods lead, necessarily, to different layer qualities.

For this thesis we have successfully grown graphene mono- and multilayers on different substrates and have measured their spectral functions respectively. Our method serves to elucidate the substrate-induced modifications to the electronic structure of graphene. Before the measured effects are summarized, a small discussion on the growth mechanisms that have been used will be given.

6.1.1 Comparison of the different growth mechanisms

Three different graphene growth mechanisms have been investigated in this thesis to obtain graphene layers:

- Growth *via* the segregation of carbon atoms from the bulk on Ru(0001) (chapter 3).

6. SUMMARY, CONCLUSIONS AND OUTLOOK

- Graphene growth *via* diffusion of nickel atoms on the SiC(0001) surface, which transforms the substrate into Θ-Ni$_2$Si (chapter 4).

- Chemical vapor deposition (CVD) on the Ni(111) surface by cracking of propylene molecules (chapter 5).

The segregation method on Ru(0001) leads to growth of homogeneous graphene layers. The method is simple and can easily be manipulated, as proved by the controlled growth of graphene multilayers. Moreover, the sharp bands in the ARPES-spectra suggest that the first layer was of exceptional quality.

Low-temperature graphene growth on SiC leads to a rather inhomogeneous graphene growth. Although such a statement should be made carefully, since no local probe experiments have been performed, it seems that our samples prepared with this method show limited quality. Besides the graphene monolayer, bilayer islands of varying orientations were achieved.

CVD on Ni(111) leads to excellent graphene layers, as also known from previous studies [37, 156]. The lower electron density on top of the graphene layer limits the growth to exactly one monolayer, and the matching lattice constant of the Ni(111) surface to that of graphene makes this method particularly suitable for the growth of graphene wafers.

6.1.2 Comparison of the interaction of graphene with the different substrates

The graphene layers on the different substrates showed explicitly different characteristics.

Graphene on Ru(0001) does not exhibit Dirac-like behavior of the charge carriers in the first graphene layer. The graphene π-bands are hybridized with the Ru $4d$-bands and the graphene-typical linearly dispersive electronic structure close to the Fermi level is not preserved. However, the second graphene layer behaves, basically, like freestanding graphene with slight electron doping. The ARPES-spectra for this system shows electron-plasmon- and electron-phonon-interaction induced kinks and no band gap at the Dirac-point. An additional graphene layer results in a gapped bilayer-like system, similar to the well-known respective interlayer system on SiC(0001).

Graphene on Au on Ru(0001) has a spectral function that is similar to free-standing graphene with a 200 meV band gap at the former Dirac-point. This band gap results from the weak interaction of the graphene p_z-orbitals with the underlying gold atoms and the special lattice mismatch, which results in a sublattice symmetry breaking. However, the band gap lies about 150 meV above the Fermi level since gold induces hole-doping in the graphene layer, and could only be made visible in the ARPES spectra by potassium-induced electron-doping.

Graphene on Ni(111) and on θ-Ni$_2$Si shows strong hybridization between Ni $3d$ and graphene π valence band states. Both systems are very similar in terms of spectral function, although the Ni d-bands do not cross the Fermi level for Θ-Ni$_2$Si. The lowered electron density atop the graphene layer makes the graphene-growth on Ni(111) self-limiting to one monolayer, while on Θ-Ni$_2$Si bilayer islands of different orientations appear atop the first layer. For the graphene/Ni(111)-system we showed many hybridization effects that have been theoretically predicted, but not previously observed.

6.2 Conclusions

In spite of the fact that the segregation and the CVD-method produced good-quality graphene layers, while the Ni-diffusion method of graphene growth on SiC did not lead to such high quality results, further conclusions should be drawn with regards to possible future applications.

6.2.1 Possible Future Applications

The CVD-method of graphene growth on nickel has already led to graphene layers of extraordinary size and quality [37]. Therefore, it seems likely that future graphene wafers will be produced on Ni(111) *via* CVD[1]. However, most electronic applications

---

[1] Although graphene wafers of similar size and quality have also already been grown on Cu-substrates [44].

6. SUMMARY, CONCLUSIONS AND OUTLOOK

would require either free-standing graphene, or graphene layers that are situated on a semi-conductor, which means that the graphene layers have to be separated from these substrates, e.g. by Kim *et al.* [37].

Moreover, as discussed in Chapter 3, free-standing graphene exhibits no band-gap. Although it is in principle possible to build most binary devices on a low/high-current basis, this is far more complicated than using simple doping-induced metal/insulator transitions. We have shown in Chapter 3 that it is in principal possible to create a clear substrate-induced band gap, but gold, as a field-effect transistor substrate for future applications is rather unlikely, gold being a metal. Due to the fact that the Dirac characteristics of the charge carriers in graphene are automatically destroyed by the appearance of a band gap at the Dirac-point, it is also questionable whether a technical approach towards gapped graphene-based devices is the most promising.

On nickel and Θ-Ni_2Si, the graphene π-bands are strongly hybridized and the charge carriers do not exhibit Dirac-behavior. This makes such a system unsuitable for graphene-based field-effect transistors, but other applications might be possible.

6.2.1.1 Graphene as a spin filter

Graphene on nickel could be used as a spin filter in future spintronics devices. However, the spin filtering is better with graphene multilayers. Such layers probably cannot be grown by CVD, due to the low binding energy of the second layer, but other methods could lead to perfect spin filters.

6.3 Outlook

The number of graphene publications is still increasing exponentially [4], and it is highly probable that graphene will enter every-day electronics within a few decades. Besides possible future electronic applications that have already been discussed in this thesis, graphene's stability and optical properties also make it a promising candidate for flexible electronics and touch screens [44].

Moreover, graphene provides a perfect laboratory as a material for extensive fundamental research. There are still many possible substrates that have not been studied *via* angular-resolved photoemission spectroscopy to clarify the modifications of the graphene band structure. Whatever graphene research will bring to us, it might not only change our everyday life, but also our understanding of fundamental physics.

6. SUMMARY, CONCLUSIONS AND OUTLOOK

Bibliography

[1] A. K. Geim, "Graphene: Status and Prospects," *Science*, vol. 324, no. 5934, pp. 1530–1534, 2009. 1, 11

[2] A. H. C. Neto, F. Guinea, N. M. R. Peres, K. S. Novoselov, and A. K. Geim, "The electronic properties of graphene," *Reviews of Modern Physics*, vol. 81, pp. 109–161, 2009. 1, 11

[3] D. R. Cooper, B. DAnjou, and N. Ghattamaneni, "Experimental review of graphene," *ISRN Condensed Matter Physics*, vol. 2012, p. 501686, 2012. 1

[4] A. Barth and W. Marx, "Graphene - a rising star in view of scientometrics," Tech. Rep. arXiv:0808.3320, Aug 2008. 1, 118

[5] S. Wang and P. R. Buseck, "Packing of C_{60} molecules and related fullerenes in crystals: a direct view," *Chemical Physics Letters*, vol. 182, pp. 1–4, jul 1991. 2

[6] H. Kroto, J. Heath, S. O'Brien, R. Curl, and R. Smalley, "C_{60}: Buckminsterfullerene," *Nature*, vol. 318, pp. 162–163, 1985. 2

[7] H.-P. Boehm, R. Setton, and E. Stumpp, "Nomenclature and terminology of graphite intercalation compounds (IUPAC Recommendations 1994)," *Pure Appl. Chem.*, vol. 66, no. 9, pp. 1893–1901, 1994. 1

[8] R. Saito, G. Dresselhaus, and M. S. Dresselhaus, *Physical Properties of Carbon Nanotubes*. Imperial College Press, 1998. 1, 5, 7, 19

[9] P. R. Wallace, "The band theory of graphite," *Phys. Rev.*, vol. 71, pp. 622–634, May 1947. 1, 3, 5

BIBLIOGRAPHY

[10] G. S. Painter and D. E. Ellis, "Electronic band structure and optical properties of graphite from a variational approach," *Phys. Rev. B*, vol. 1, pp. 4747–4752, Jun 1970. 1

[11] J. W. McClure, "Band structure of graphite and de haas-van alphen effect," *Phys. Rev.*, vol. 108, pp. 612–618, Nov 1957. 3

[12] J. C. Slonczewski and P. R. Weiss, "Band structure of graphite," *Phys. Rev.*, vol. 109, pp. 272–279, Jan 1958. 3

[13] D. P. DiVincenzo and E. J. Mele, "Self-consistent effective-mass theory for intralayer screening in graphite intercalation compounds," *Phys. Rev. B*, vol. 29, no. 4, pp. 1685–94, 1984. 3, 9, 10

[14] L. D. Landau, "Zur Theorie der Phasenumwandlungen II.," *Phys. Z. Sowjetunion*, vol. 11, p. 2635. 3

[15] A. J. van Bommel, J. E. Crombeen, and A. van Tooren, "LEED and Auger Electron Observations of the SiC(0001) Surface," *Surface Science*, vol. 48, pp. 463–472, 1974. 3, 12, 77

[16] C. OSHIMA, E. BANNAI, T. TANAKA, and S. KAWAI, "Carbon Layer on Lanthanum Hexaboride (100) Surface," *Japanese journal of applied physics*, vol. 16, no. 6, pp. 965–969, 19770605. 3

[17] F. Himpsel, K. Christmann, P. Heimann, D. Eastman, and P. J. Feibelman, "Adsorbate band dispersions for C on Ru(0001)," *Surface Science*, vol. 115, no. 3, pp. L159 – L164, 1982. 3, 15, 62

[18] R. Rosei, M. De Crescenzi, F. Sette, C. Quaresima, A. Savoia, and P. Perfetti, "Structure of graphitic carbon on Ni(111): A surface extended-energy-loss finestructure study," *Phys. Rev. B*, vol. 28, pp. 1161–1164, Jul 1983. 3, 15, 97

[19] N. Kholin, E. Rut'kov, and A. Tontegode, "The nature of the adsorption bond between graphite islands and iridium surface," *Surface Science*, vol. 139, no. 1, pp. 155 – 172, 1984. 3

BIBLIOGRAPHY

[20] K. S. Novoselov, A. K. Geim, S. V. Morosov, D. Jiang, Y. Zhang, S. V. Dubonos, I. V. Grigorieva, and A. A. Firsov, "Electric field effect in atomically thin carbon films," *Science*, vol. 306, p. 666, 2004. 3, 4, 10, 11

[21] K. S. Novoselov, E. McCann, S. V. Morosov, V. Fal'ko, M. I. Katsnelson, U. Zeitler, D. Jiang, F. Schedin, and A. K. Geim, "Two-dimensional gas of massless dirac fermions in graphene," *Nature*, vol. 438, pp. 192–200, 2005. 3, 10, 11

[22] K. S. Novoselov, D. Jiang, F. Schedin, T. J. Booth, V. V. Khotkevich, S. V. Morozov, and A. K. Geim, "Two-dimensional atomic crystals," *Proc. Nat. Acad. Sci.*, vol. 102, no. 30, pp. 10451–10453, 2005. 3, 10

[23] C. Berger, Z. Song, T. Li, X. Li, A. Y. Ogbazghi, R. Feng, Z. Dai, A. N. Marchenkov, E. H. Conrad, P. N. First, and W. A. deHeer, "Ultrathin Epitaxial Graphite: 2D Electron Gas Properties and a Route toward Graphene-based Nanoelectronics," *J. Phys. Chem. B*, vol. 108, no. 52, pp. 19912–19916, 2004. 4, 11, 12, 77, 88

[24] Y. Zhang, Y. W. Tan, H. L. Stormer, and P. Kim, "Experimental observation of the quantum hall effect and berry's phase in graphene," *Nature*, vol. 438, pp. 201–204, 2005. 4, 10

[25] A. Bostwick, T. Ohta, J. L. McChesney, K. V. Emtsev, T. Seyller, K. Horn, and E. Rotenberg, "Symmetry breaking in few layer graphene films," *New J. Phys*, vol. 9, p. 385, 2007. 9, 13, 14, 57, 62, 63, 65, 66, 72, 73, 75

[26] C. Cohen-Tannoudji, B. Diu, and F. Laloë, *Mechanique Quantique*. 1977. 9

[27] E. Rebhan, *Theoretische Physik ll.* 2004. 9

[28] C. Berger, Z. Song, X. Li, X. Wu, N. Brown, C. Naud, D. Mayou, T. Li, J. Hass, A. Marchenkov, E. Conrad, P. N. First, and W. A. de Heer, "Electronic confinement and coherence in patterned epitaxial graphene," *Science*, vol. 312, pp. 1191–6, 2006. 10, 14, 77

BIBLIOGRAPHY

[29] F. Schedin, A. K. Geim, S. V. Morozov, D. Jiang, E. H. Hill, P. Blake, and K. S. Novoselov, "Detection of individual gas molecules absorbed on graphene," *Nature Materials*, vol. 6, p. 652, 2007. 10

[30] K. S. Novoselov, Z. Jiang, Y. Zhang, S. V. Morozov, H. L. Stormer, U. Zeitler, J. C. Maan, G. S. Boebinger, P. Kim, and A. K. Geim, "Room-Temperature Quantum Hall Effect in Graphene," *Science*, vol. 315, no. 5817, pp. 1379–, 2007. 10

[31] A. Shapere and F. Wilczek, *Geometric phases in physics*. World Scientific, 1976. 10

[32] M. I. Katsnelson, "Zitterbewegung, chirality, and minimal conductivity in graphene," *The European Physical Journal B*, vol. 51, p. 157, 2006. 11

[33] H. Zhang, C.-X. Liu, X.-L. Qi, X. Dai, Z. Fang, and S.-C. Zhang, "Topological insulators in bi2se3, bi2te3 and sb2te3 with a single dirac cone on the surface," *Nature Physics*, vol. 5, pp. 438–442. 11

[34] A. K. Geim and K. S. Novoselov, "The rise of graphene," *Nat. Mater.*, vol. 6, no. 3, pp. 183–191, 2007. 11, 57

[35] Z. Chen, Y.-M. Lin, M. J. Rooks, and P. Avouris, "Graphene nano-ribbon electronics," *Physica E: Low-dimensional Systems and Nanostructures*, vol. 40, no. 2, pp. 228 – 232, 2007. International Symposium on Nanometer-Scale Quantum Physics. 11

[36] W. Zhu, V. Perebeinos, M. Freitag, and P. Avouris, "Carrier scattering, mobilities and electrostatic potential in mono-, bi- and tri-layer graphenes," 2009. 11

[37] K. S. Kim, Y. Zhao, H. Jang, S. Y. Lee, J. M. Kim, K. S. Kim, J.-H. Ahn, P. Kim, J.-Y. Choi, and B. H. Hong, "Large-scale pattern growth of graphene films for stretchable transparent electrodes," *Nature*, 2009. 11, 97, 116, 117, 118

[38] "Reduced graphene oxide doped with dopant, thin layer and transparent electrode," *United States Patent Application*. 11

BIBLIOGRAPHY

[39] B. Wang, M.-L. Bocquet, S. Marchini, S. Gnther, and J. Wintterlin, "Chemical origin of a graphene moiré overlayer on Ru(0001)," *Phys. Chem. Chem. Phys.*, vol. 10, no. 24, pp. 3530–3534, 2008. 11, 60, 62

[40] A. T. Tilke, F. C. Simmel, R. H. Blick, H. Lorenz, and J. P. Kotthaus, "Coulomb blockade in silicon nanostructures," *Progress in Quantum Electronics*, vol. 25, no. 3, pp. 97–138, 2001. 11

[41] Y. Takahashi, Y. Ono, A. Fujiwara, and H. Inokawa, "Silicon single-electron devices," *Journal of Physics: Condensed Matter*, vol. 14, no. 39, p. R995, 2002. 11

[42] V. M. Karpan, G. Giovanneti, P. A. Khomyakov, M. Talanana, A. A. Starikov, M. Zwierzycki, J. van den Brink, G. Brocks, and P. J. Kelly, "Graphene and graphite as perfect spin filters," *Phys. Rev. Lett.*, vol. 99, p. 176602, 2007. 11, 71, 87, 98, 99, 100, 101, 105, 106, 107, 108, 109, 112

[43] Y. S. Dedkov, M. Fonin, and C. Laubschat, "A possible source of spin-polarized electrons: The inert graphene/Ni(111) system," *Applied Physics Letters*, vol. 92, no. 5, p. 052506, 2008. 11, 105, 107

[44] S. Bae, H. K. Kim, X. Xu, J. Balakrishnan, T. Lei, Y. I. Song, Y. J. Kim, B. Ozyilmaz, J.-H. Ahn, B. H. Hong, and S. Iijima, "30-inch roll-based production of high-quality graphene films for flexible transparent electrodes," 2009. 11, 117, 118

[45] I. Forbeaux, J. M. Themlin, and J. M. Debever, "Heteroepitaxial graphite on 6H-SiC(0001): Interface formation through conduction-band electronic structure," *Phys. Rev. B*, vol. 58, no. 24, pp. 16396–406, 1998. 12

[46] I. Forbeaux, J. M. Themlin, and J. M. Debever, "High-temperature graphitization of the 6H-SiC(0001) face," *Surf. Science*, vol. 442, no. 1, pp. 9–18, 1999. 12, 77

[47] L. Johansson, F. Owman, and P. Martensson, "High-resolution core level study of 6H-SiC(0001)," *Phys. Rev. B*, vol. 53, no. 20, pp. 13793–13802, 1995. 12, 82, 83

BIBLIOGRAPHY

[48] P. Martensson, F. Owman, and L. I. Johansson, "Morphology, atomic and electronic structure of 6H-SiC(0001) surfaces," *phys. stat. sol.*, vol. 202, pp. 501–528, 1997. 12, 82, 83, 84

[49] K. V. Emtsev, T. Seyller, F. Speck, L. Ley, P. Stojanov, J. D. Riley, and R. G. C. Leckey, "Initial stages of the graphite-SiC(0001) interface formation studied by photoelectron spectroscopy," *cond-mat*, p. 0609383, 2006. 12, 88

[50] A. Bostwick, T. Ohta, T. Seyller, K. Horn, and E. Rotenberg, "Quasiparticle dynamics in graphene," *Nat. Phys.*, vol. 3, no. 1, pp. 36–40, 2007. 12, 13, 14, 23, 57, 62, 65, 66, 69, 73, 75, 77, 88

[51] T. Ohta, F. E. Gabaly, A. Bostwick, J. L. McChesney, K. V. Emtsev, A. K. Schmid, T. Seyller, K. Horn, and E. Rotenberg, "Morphology of graphene thin film growth on SiC(0001)," *New Journal of Physics*, vol. 10, p. 023034, 2008. 12, 77, 94

[52] U. Starke, C. Bram, P.-R. Steiner, W. Hartner, L. Hammer, K. Heinz, and K. Müller, "The (0001)-surface of 6H-SiC: morphology, composition and structure," *Applied Surface Science*, vol. 89, pp. 175 – 185, 1995. 12, 82

[53] K. V. Emtsev, A. Bostwick, K. Horn, J. Jobst, G. L. Kellogg, L. Ley, J. L. McChesney, T. Ohta, S. A. Reshanov, J. Rohrl, E. Rotenberg, A. K. Schmid, D. Waldmann, H. B. Weber, and T. Seyller, "Towards wafer-size graphene layers by atmospheric pressure graphitization of silicon carbide," *Nat. Mater.*, vol. 8, no. 3, pp. 203–207, 2009. 12, 77

[54] C. Virojanadra, M. Syvarjarvi, R. Yakimova, L. I. Johansson, A. A. Zakharov, and T. Balasubramanian, "Homogeneous large-area graphene layer growth on 6H-SiC(0001)," *Phys. Rev. B*, vol. 178, p. 245403, 2008. 12, 77

[55] J. Hass, F. Varchon, J. E. Millan-Otoya, M. Sprinkle, N. Sharma, W. A. de Heer, C. Berger, P. N. First, L. Magaud, and C. E. H., "Why multilayer graphene on 4H-SiC(000$\bar{1}$) behaves like a single sheet of graphene," *Phys. Rev. Lett.*, vol. 100, p. 125504, 2008. 12, 14, 77, 90

BIBLIOGRAPHY

[56] K. V. Emtsev, F. Speck, T. Seyller, L. Ley, and J. D. Riley, "Interaction, growth, and ordering of epitaxial graphene on SiC(0001) surfaces: A comparative photoelectron spectroscopy study," *Phys. Rev. B*, vol. 77, no. 15, pp. 155303–10, 2008. 13, 73, 77, 84

[57] C. Enderlein, Y. S. Kim, A. Bostwick, E. Rotenberg, and K. Horn, "The formation of an energy gap in graphene on ruthenium by controlling the interface," *New Journal of Physics*, vol. 12, no. 3, 2010. 13, 14, 58, 73

[58] S. Y. Zhou, G. H. Gweon, A. V. Fedorov, P. N. First, W. A. de Heer, D. H. Lee, F. Guinea, A. H. Castro Neto, and A. Lanzara, "Substrate-induced bandgap opening in epitaxial graphene," *Nat. Mater.*, vol. 6, pp. 770–775, 2007. 13, 14, 57

[59] S. Y. Zhou, D. A. Siegel, A. V. Federov, F. El Gabaly, A. K. Schmid, A. H. Castro Neto, D. H. Lee, and A. Lanzara, "Origin of the energy bandgap in epitaxial graphene [reply]," *Nat. Mater.*, 2008. 14, 58

[60] E. Rotenberg, A. Bostwick, T. Ohta, J. L. McChesney, T. Seyller, and K. Horn, "Origin of the energy bandgap in epitaxial graphene," *Nat. Mater.*, vol. 7, pp. 258–259, 2008. 14, 58, 93

[61] I. Forbeaux, J. M. Themlin, A. Charrier, F. Thibaudau, and J. M. Debever, "Solid-state graphitization mechanisms of silicon carbide 6h-sic polar faces," *Applied Surface Science*, vol. 162-163, pp. 406 – 412, 2000. 14

[62] X. Wu, X. Li, Z. Song, C. Berger, and W. A. de Heer, "Weak antilocalization in epitaxial graphene: Evidence for chiral electrons," *Phys. Rev. Lett.*, vol. 98, no. 13, pp. 136801–4, 2007. 14

[63] I. Pletikosić, M. Kralj, P. Pervan, R. Brako, J. Coraux, A. T. N'Diaye, C. Busse, and T. Michely, "Dirac Cones and Minigaps for Graphene on Ir(111)," *Phys. Rev. Lett.*, vol. 102, p. 056808, Feb 2009. 15

[64] J. Shelton, H. R. Patil, and J. M. Blakely, "Equilibrium segregation of carbon to a nickel (111) surface: A surface phase transition," *Surface Science*, vol. 43, pp. 493–520, June 1974. 15, 97

BIBLIOGRAPHY

[65] A. Nagashima, N. Tejima, and C. Oshima, "Electronic states of the pristine and alkali-metal-intercalated monolayer graphite/Ni(111) systems," *Phys. Rev. B*, vol. 50, pp. 17487–17495, Dec 1994. 15, 97, 98, 99, 105

[66] P. Sutter, M. S. Hybertsen, J. T. Sadowski, and E. Sutter, "Electronic Structure of Few-Layer Epitaxial Graphene on Ru(0001)," *Nano Lett.*, vol. 9, no. 7, pp. 2654–2660, 2009. 15, 58, 60, 61, 62, 63, 73

[67] A. Grüneis and D. V. Vyalikh, "Tunable hybridization between electronic states of graphene and a metal surface," *Phys. Rev. B*, vol. 77, no. 19, p. 193401, 2008. 15, 72, 87, 97, 98, 105

[68] H. Madden and G. Ertl, "Decomposition of carbon monoxide on a (110) nickel surface," *Surface Science*, vol. 35, pp. 211 – 226, 1973". 15, 97

[69] A. E. Becquerel, "Mémoire sur les effets électriques produits sous l'influence des rayons solaires," *Comptes Rendus des Séances Hebdomadaires*, vol. 9, pp. 561–567, 1839. 17

[70] H. Hertz, "Über einen Einfluss des ultravioletten Lichtes auf die elektrische Entladung," *Annalen der Physik und Chemie*, vol. 31, 1887. 17

[71] H. Hertz, *Untersuchungen ber die Ausbreitung der elektrischen Kraft*. 1892. 17

[72] P. Lenard, "Ueber die lichtelektrische wirkung," *Annalen der Physik*, vol. 313, pp. 149–198, 1902. 17

[73] A. Einstein, "Über einen die Erzeugung und Verwandlung des Lichtes betreffenden heuristischen Gesichtspunkt," *Annalen der Physik*, vol. 322, pp. 132–148, 1905. 17

[74] W. E. Spicer, "Photoemissive, photoconductive, and optical absorption studies of alkali-antimony compounds," *Phys. Rev.*, vol. 112, pp. 114–122, Oct 1958. 18

[75] C. N. Berglund and W. E. Spicer, "Photoemission studies of copper and silver: Theory," *Phys. Rev.*, vol. 136, pp. A1030–A1044, Nov 1964. 18

[76] T. N. Rhodin and J. Gadzuk, *The nature of the surface chemical bond*. 1979. 21

BIBLIOGRAPHY

[77] G. Somorjai, *Chemistry in two dimensions: surfaces*. 1981. 21

[78] D. R. Penn, "Electron mean free paths for free-electron-like materials," *Phys. Rev. B*, vol. 13, pp. 5248–5254, Jun 1976. 21

[79] H. Matsui, K. Terashima, T. Sato, T. Takahashi, S. C. Wang, H. B. Yang, H. Ding, T. Uefuji, and K. Yamada, "Angle-Resolved Photoemission Spectroscopy of the Antiferromagnetic Superconductor Nd1.87Ce0.13CuO4: Anisotropic Spin-Correlation Gap, Pseudogap, and the Induced Quasiparticle Mass Enhancement," *Phys. Rev. Lett.*, vol. 94, p. 047005, 2005. 23

[80] J. Mesot, M. R. Norman, H. Ding, M. Randeria, J. C. Campuzano, A. Paramekanti, H. M. Fretwell, A. Kaminski, T. Takeuchi, T. Yokoya, T. Sato, T. Takahashi, T. Mochiku, and K. Kadowaki, "Superconducting gap anisotropy and quasiparticle interactions: A doping dependent photoemission study," *Phys. Rev. Lett.*, vol. 83, pp. 840–843, Jul 1999. 23

[81] A. Bostwick, T. Ohta, J. L. McChesney, T. Seyller, K. Horn, and E. Rotenberg, "Renormalization of graphene bands by many-body interactions," *Solid State Communications*, vol. 143, pp. 63–71, 2007. 23, 62, 63, 69

[82] T. Valla, A. Fedorov, P. Johnson, and S. Hulbert, "Many-body effects in angle-resolved photoemission: quasiparticle energy and lifetime of a Mo(110) suface state," *Phys. Rev. Lett.*, vol. 83, pp. 2085–8, 1999. 23

[83] T. Pelzer, G. Ceballos, F. Zbikowski, B. Willerding, K. Wandelt, T. U., C. Reu, T. Fauster, and J. Braun, "Electronic structure of the Ru(0001) surface," *Journal of Physics: Condensed Matter*, vol. 12, no. 10, 2000. 25, 61

[84] M. Z. Hasan, Y.-D. Chuang, D. Qian, Y. W. Li, Y. Kong, A. Kuprin, A. V. Fedorov, R. Kimmerling, E. Rotenberg, K. Rossnagel, Z. Hussain, H. Koh, N. S. Rogado, M. L. Foo, and R. J. Cava, "Fermi Surface and Quasiparticle Dynamics of Na0.7CoO2 Investigated by Angle-Resolved Photoemission Spectroscopy," *Phys. Rev. Lett.*, vol. 92, p. 246402, Jun 2004. 25

[85] E. Grannemann and M. van der Wiel, *Handbook on synchrotron radiation*. 1983. 28

BIBLIOGRAPHY

[86] C. Davisson and L. H. Germer, "Diffraction of electrons by a crystal of nickel," *Phys. Rev.*, vol. 30, pp. 705–740, Dec 1927. 34

[87] H. Kramers, "On the theory of x-ray absorption and of the continuous x-ray spectrum," *Philosophical Magazine*, vol. 46, pp. 836–871, november 1923. 34

[88] T. Braun, *Entwicklung und Charakterisierung einer Elektronenlinse für abbildende Photoelektronenspektroskopie.* Berlin: Freie Universität, Jan 2009. $L_E DOC$. 40

[89] H. Jones, "Applications of the bloch theory to the study of alloys and of the properties of bismuth," *Proceedings of the Royal Society of London. Series A, Mathematical and Physical Sciences (1934-1990)*, vol. 147, no. 861, pp. 396–417, 1934. 57, 63

[90] M. Ezawa, "Peculiar width dependence of the electronic properties of carbon nanoribbons," *Phys. Rev. B*, vol. 73, p. 045432, Jan 2006. 57

[91] K. Nakada, M. Fujita, G. Dresselhaus, and M. S. Dresselhaus, "Edge state in graphene ribbons: Nanometer size effect and edge shape dependence," *Phys. Rev. B*, vol. 54, no. 24, p. 17954, 1996. 57

[92] E. McCann and V. Fal'ko, "Landau-level degeneracy and quantum hall effect in a graphite bilayer," *Phys. Rev. Lett.*, vol. 96, p. 086805, 2006. 57

[93] T. Ohta, A. Bostwick, T. Seyller, K. Horn, and E. Rotenberg, "Controlling the electronic structure of bilayer graphene," *Science*, vol. 313, pp. 951–954, 2006. 57, 62, 65, 66

[94] G. Giovannetti, P. A. Khomyakov, G. Brocks, V. M. Karpan, J. van den Brink, and P. J. Kelly, "Doping graphene with metal contacts," *Phys. Rev. Lett.*, vol. 101, no. 2, pp. 026803–4, 2008. 57, 72

[95] E. H. Hwang, B. Yu-Kuang Hu, and S. Das Sarma, "Inelastic carrier lifetime in graphene," *Phys. Rev. B*, vol. 76, p. 115434, 2007. 57

[96] M. Polini, R. Asgari, G. Borghi, Y. Barlas, T. Pereg-Barnea, and A. H. MacDonald, "Plasmons and the spectral function of graphene," *Phys. Rev. B*, vol. 77, no. 8, pp. 081411–4, 2008. 57, 69

BIBLIOGRAPHY

[97] E. H. Hwang and S. D. Sarma, "Quasiparticle spectral function in doped graphene: Electron-electron interaction effects in ARPES," *Phys. Rev. B*, vol. 77, no. 8, pp. 081412–4, 2008. 57, 69

[98] C.-H. Park, F. Giustino, C. D. Spataru, M. L. Cohen, and S. G. Louie, "First-principles study of electron linewidths in graphene," *Phys. Rev. Lett.*, vol. 102, no. 7, pp. 076803–4, 2009. 57

[99] C.-H. Park, F. Giustino, C. D. Spataru, M. L. Cohen, and S. G. Louie, "Angle-resolved photoemission spectra of graphene from first-principles calculations," *Nano Lett.*, vol. 9, pp. 4234–4239, 2009. 57, 69

[100] S. Kim, J. Ihm, H. J. Choi, and Y.-W. Son, "Origin of anomalous electronic structures of epitaxial graphene on silicon carbide," *Phys. Rev. Lett.*, vol. 100, no. 17, pp. 176802–4, 2008. 57

[101] A. Bostwick, J. L. McChesney, K. V. Emtsev, T. Seyller, K. Horn, S. D. Kevan, and E. Rotenberg, "Quasiparticle transformation during a metal insulator transition in graphene," *Phys. Rev. Lett.*, vol. 103, no. 5, pp. 056404-1–4, 2009. 58, 77

[102] P. W. Sutter, J.-I. Flege, and E. A. Sutter, "Epitaxial graphene on ruthenium," *Nat. Mater.*, vol. 7, no. 5, pp. 406–411, 2008. 58, 60, 75

[103] A. L. V. de Parga, F. Calleja, B. Borca, J. M. C. G. Passeggi, J. J. Hinarejos, F. Guinea, and R. Miranda, "Periodically rippled graphene: Growth and spatially resolved electronic structure," *Phys. Rev. Lett.*, vol. 100, no. 5, pp. 056807–4, 2008. 58, 60

[104] S. Marchini, S. Günther, and J. Wintterlin, "Scanning tunneling microscopy of graphene on Ru(0001)," *Phys. Rev. B*, vol. 76, p. 075429, Aug 2007. 58

[105] D. Martoccia, P. R. Willmott, T. Brugger, M. Bjorck, S. Gunther, C. M. Schleputz, A. Cervellino, S. A. Pauli, B. D. Patterson, S. Marchini, J. Wintterlin, W. Moritz, and T. Greber, "Graphene on Ru(0001): A 25 x 25 Supercell," *Phys. Rev. Lett.*, vol. 101, no. 12, pp. 126102–4, 2008. 60

BIBLIOGRAPHY

[106] D.-E. Jiang, M.-H. Du, and S. Dai, "First principles study of the graphene/Ru(0001) interface," *J. Chem. Phys.*, vol. 130, no. 7, pp. 074705–5, 2009. 60

[107] T. Brugger, S. Gunther, B. Wang, J. H. Dil, M.-L. Bocquet, J. Osterwalder, J. Wintterlin, and T. Greber, "Comparison of electronic structure and template function of single-layer graphene and a hexagonal boron nitride nanomesh on Ru(0001)," *Phys. Rev. B*, vol. 79, no. 4, pp. 045407–6, 2009. 60, 62

[108] M. Lindroos, P. Hofmann, and D. Menzel, "Angle-resolved photoemission determination of the band structure of Ru(001)," *Phys. Rev. B*, vol. 33, pp. 6798–6809, May 1986. 61

[109] T. Ohta, A. Bostwick, J. L. McChesney, T. Seyller, K. Horn, and E. Rotenberg, "Interlayer interaction and electronic screening in multilayer graphene investigated with angle-resolved photoemission spectroscopy," *Phys. Rev. Lett.*, vol. 98, no. 20, pp. 206802–1–4, 2007. 62, 77

[110] Y. S. Dedkov, A. M. Shikin, V. K. Adamchuk, S. L. Molodtsov, C. Laubschat, A. Bauer, and G. Kaindl, "Intercalation of copper underneath a monolayer of graphite on Ni(111)," *Phys. Rev. B*, vol. 64, no. 3, p. 035405, 2001. 62, 100

[111] A. Varykhalov, J. Sanchez-Barriga, A. M. Shikin, C. Biswas, E. Vescovo, A. Rybkin, D. Marchenko, and O. Rader, "Electronic and magnetic properties of quasifreestanding graphene on Ni," *Phys. Rev. Lett.*, vol. 101, no. 15, pp. 157601–4, 2008. 62, 100

[112] E. Shirley, L. Terminello, A. Santoni, and F. J. Himpsel, "Brillouin-zone-selection effects in graphite photoelectron angular distributions," *Phys. Rev. B*, vol. 51, no. 19, pp. 13614–22, 1995. 63, 72

[113] A. M. Shikin, G. V. Prudnikova, V. K. Adamchuk, F. Moresco, and K. H. Rieder, "Surface intercalation of gold underneath a graphite monolayer on Ni(111) studied by angle-resolved photoemission and high-resolution electron-energy-loss spectroscopy," *Phys. Rev. B*, vol. 62, no. 19, p. 13202, 2000. 66

BIBLIOGRAPHY

[114] I. Gierz, C. Riedl, U. Starke, C. R. Ast, and K. Kern, "Atomic hole doping of graphene," *Nano Letters*, vol. 8, no. 12, pp. 4603–4607, 2008. 66

[115] A. Varykhalov, M. R. Scholz, T. K. Kim, and O. Rader, "Effect of noble-metal contacts on doping and band gap of graphene," 2010. 71

[116] M. Pivetta, F. Patthey, I. Barke, H. Hovel, B. Delley, and W.-D. Schneider, "Gap opening in the surface electronic structure of graphite induced by adsorption of alkali atoms: Photoemission experiments and density functional calculations," *Phys. Rev. B*, vol. 71, no. 16, pp. 165430–4, 2005. 71

[117] M. Mucha-Kruczynski, O. Tsyplyatyev, A. Grishin, E. McCann, V. Fal'ko, A. Bostwick, and E. Rotenberg, "Characterization of graphene through anisotropy of constant-energy maps in angle-resolved photoemission," *Phys. Rev. B*, vol. 77, p. 195403, 2008. 72

[118] M. Naitoh, M. Kitada, S. Nishigaki, N. Toyama, and S. F., "An STM observation of the initial process of graphitization at the 6H-SiC(000$\bar{1}$) surface," *Surf. Rev. Lett.*, vol. 10, p. 473, 2003. 77

[119] J. Hass, R. Feng, T. Li, Z. Zong, W. A. de Heer, P. N. First, E. H. Conrad, C. A. Jeffrey, and C. Berger, "Highly ordered graphene for two dimensional electronics," *Appl. Phys. Lett.*, vol. 89, p. 143106, 2006. 77

[120] K. Robbie, S. Jemander, N. Lin, C. Hallin, R. Erlandsson, G. V. Hansson, and M. L.D., "Formation on Ni-graphite intercalation compounds on SiC," *Phys. Rev. B*, vol. 64, p. 155401, 2001. 77, 78, 85

[121] Z.-Y. Juang, C.-Y. Wu, C.-W. Lo, W.-Y. Chen, C.-F. Huang, J.-C. Hwang, F.-R. Chen, K.-C. Leou, and C.-H. Tsai, "Synthesis of graphene on silicon carbide substrates at low temperature," *Carbon*, vol. 47, pp. 2026–2031, 2009. 77, 78, 79

[122] A. Hähnel, E. Pippel, V. Ischenko, and J. Woltersdorf, "Nanostructuring in Ni/SiC reaction layers, investigated by imaging of atomic columns and DFT calculations," *Materials Chemestry and Physics*, vol. 114, pp. 802 – 808, 2009. 77, 78, 79, 88

BIBLIOGRAPHY

[123] T. Fujimura and S.-I. Tanaka, "In-situ high temperature X-ray diffraction study of Ni/SiC interface reactions," *Journal of Materials Science*, vol. 34, pp. 235 – 239, 1999. 78, 88

[124] F. Goesmann and R. Schmid-Fetzer, "Metals on 6H-SiC: contact formation from the materials science point of view," *Materials Science and Engineering B*, vol. 46, no. 1-3, pp. 357 – 362, 1997. E-MRS 1996 Spring Meeting, Symposium A: High Temperature Electronics: Materials, Devices and Applications. 77, 88

[125] I. P. Nikitina, K. V. Vassilevski, N. G. Wright, A. B. Horsfall, A. O'Neill, and C. M. Johnson, "Formation and role of graphite and nickel silicide in nickel based ohmic contacts to n-type silicon carbide," *Journal of Applied Physics*, vol. 97, p. 083709, 2005. 77

[126] F. La Via, F. Roccaforte, A. Makhtari, V. Raineri, P. Musumeci, and L. Calcagno, "Structural and electrical characterisation of titanium and nickel silicide contacts on silicon carbide," *Microelectron. Eng.*, vol. 60, no. 1, pp. 269–282, 2002. 77, 88

[127] A. Woodworth and C. Stinespring, "Surface chemistry of Ni induced graphite formation on the 6H-SiC(0001) surface and its implications for graphene synthesis," *Carbon*, vol. 48, no. 7, pp. 1999 – 2003, 2010. 77, 94

[128] C. S. Lim, H. Nickel, A. Naoumidis, and E. Gyrmati, "Interfacial reaction and adhesion between SiC and thin sputtered nickel films," *Journal of Materials Science*, vol. 32, pp. 6567–6572, 1997. 77

[129] A. Franciosi, J. H. Weaver, and F. Schmidt, "Electronic structure of nickel silicides Ni_2Si, $NiSi$, and $NiSi_2$," *Phys. Rev. B*, vol. 26, no. 2, pp. 546 – 553, 1982. 84, 85, 87, 88

[130] D. L. Legrand, H. W. Nesbitt, and G. M. Bancroft, "X-ray photoelectron spectroscopy study of a pristine millerite (NiS) surface and the effect of air and water oxidation," *American Mineralogist*, vol. 83, pp. 1256 – 1265, 1998. 84, 85

[131] H. Nesbitt, D. Legrand, and G. Bancroft, "Interpretation of Ni2p XPS spectra for Ni conductors and Ni insulators," *Phys. Chem. Minerals*, vol. 27, pp. 357 – 366, 2000. 84, 85

BIBLIOGRAPHY

[132] K. Balasubramanian, "Relativity and chemical bonding," *Journal of Physical Chemistry*, vol. 93, no. 18, pp. 6585–6596, 1989. 85

[133] E. Kurimoto, H. Harima, T. Toda, M. Sawada, M. Iwami, and S. Nakashima, "Raman study on the Ni/SiC interface reaction," *Journal of Applied Physics*, vol. 91, no. 12, pp. 10215–10217, 2002. 88

[134] O. Bisi, C. Calandra, U. del Pennino, P. Sassaroli, and S. Valeri, "Correlation effects in valence-band spectra of nickel silicides," *Phys. Rev. B*, vol. 30, pp. 5696–5703, Nov 1984. 88

[135] D. M. Bylander, L. Kleinman, and K. Mednick, "Self-consistent energy bands and bonding of Ni_3Si," *Phys. Rev. B*, vol. 25, pp. 1090–1095, Jan 1982. 88

[136] A. Osawa and M. Okamoto, "Influence of the Si Substrate on Nickel Silicide Formed from Thin Ni Films," *Sci. Rep. Tohuku Univ.*, vol. 27, p. 326, 1939. 88

[137] K. Toman, "The structure of Ni_2Si," *Acta Cryst.*, vol. 5, pp. 329 – 331, 1951. 88, 89

[138] K. L. Peterson, J. S. Hsiao, D. R. Chopra, and T. R. Dillingham, "Calculations of the local density of states of $NiSi_2$, $NiSi$, Ni_2Si, and Ni_3Si using the Haydock recursion method," *Phys. Rev. B*, vol. 38, pp. 9511–9516, Nov 1988. 89

[139] J. Zemann, "*Crystal structures*, 2nd edition. Vol. 1 by R. W. G. Wyckoff," *Acta Crystallographica*, vol. 18, p. 139, Jan 1965. 90

[140] A. N. Kolmogorov and V. H. Crespi, "Registry-dependent interlayer potential for graphitic systems," *Phys. Rev. B*, vol. 71, p. 235415, Jun 2005. 90, 93

[141] K. Yamamoto, M. Fukushima, T. Osaka, and C. Oshima, "Charge-transfer mechanism for the (monolayer graphite)/Ni(111) system," *Phys. Rev. B*, vol. 45, pp. 11358–11361, May 1992. 93, 97, 107

[142] Y. Gamo, A. Nagashima, M. Wakabayashi, M. Terai, and C. Oshima, "Atomic structure of monolayer graphite formed on Ni(111)," *Surface Science*, vol. 374, pp. 61–64, Mar. 1997. 97, 98

BIBLIOGRAPHY

[143] G. Odahara, T. Ishikawa, S. Otani, and C. Oshima, "Self-standing graphene sheets prepared with chemical vapor deposition and chemical etching," *e-Journal of Surface Science and Nanotechnology*, vol. 7, pp. 837–840, 2009. 97

[144] G. Bertoni, L. Calmels, A. Altibelli, and V. Serin, "First-principles calculation of the electronic structure and EELS spectra at the graphene/Ni(111) interface," *Phys. Rev. B*, vol. 71, p. 075402, Feb 2005. 98, 100, 101, 104, 105, 107, 108, 112

[145] H. Kawanowa, H. Ozawa, T. Yazaki, Y. Gotoh, and R. Souda, "Structure Analysis of Monolayer Graphite on Ni(111) Surface by Li^+-Impact Collision Ion Scattering Spectroscopy," *Japanese Journal of Applied Physics*, vol. 41, no. Part 1, No. 10, pp. 6149–6152, 2002. 98

[146] O. V. Yazyev and A. Pasquarello, "Magnetoresistive junctions based on epitaxial graphene and hexagonal boron nitride," *Phys. Rev. B*, vol. 80, p. 035408, Jul 2009. 99

[147] M. N. Baibich, J. M. Broto, A. Fert, F. N. Van Dau, F. Petroff, P. Etienne, G. Creuzet, A. Friederich, and J. Chazelas, "Giant magnetoresistance of (001)Fe/(001)Cr magnetic superlattices," *Phys. Rev. Lett.*, vol. 61, pp. 2472–2475, Nov 1988. 99

[148] G. Binasch, P. Grünberg, F. Saurenbach, and W. Zinn, "Enhanced magnetoresistance in layered magnetic structures with antiferromagnetic interlayer exchange," *Phys. Rev. B*, vol. 39, pp. 4828–4830, Mar 1989. 99

[149] C. Chappert, A. Fert, and F. N. Van Dau, "The emergence of spin electronics in data storage," *Nature Materials*, vol. 6, pp. 813–823, 2007. 99

[150] S. Yuasa, T. Nagahama, A. Fukushima, Y. Suzuki, and K. Ando, "Giant room-temperature magnetoresistance in single-crystal Fe/MgO/Fe magnetic tunnel junctions," *Nature Materials*, vol. 3, pp. 868–871, dec 2004. 99

[151] C. Heiliger, P. Zahn, B. Y. Yavorsky, and I. Mertig, "Interface structure and bias dependence of Fe|MgO|Fe tunnel junctions: Ab initio calculations," *Phys. Rev. B*, vol. 73, p. 214441, Jun 2006. 99

BIBLIOGRAPHY

[152] P. A. Dowben and R. Skomski, "Are half-metallic ferromagnets half metals? (invited)," *Journal of Applied Physics*, vol. 95, pp. 7453–7458, June 2004. 99

[153] A. Obraztsov, E. Obraztsova, A. Tyurnina, and A. Zolotukhin, "Chemical vapor deposition of thin graphite films of nanometer thickness," *Carbon*, vol. 45, no. 10, pp. 2017 – 2021, 2007. 99

[154] A. Reina, X. Jia, J. Ho, D. Nezich, H. Son, V. Bulovic, M. S. Dresselhaus, and J. Kong, "Large area, few-layer graphene films on arbitrary substrates by chemical vapor deposition," *Nano Letters*, vol. 9, pp. 30–35, Jan. 2009. 99

[155] Y. S. Dedkov, M. Fonin, U. Rüdiger, and C. Laubschat, "Rashba effect in the Graphene/Ni(111) system," *Phys. Rev. Lett.*, vol. 100, p. 107602, Mar 2008. 100, 105

[156] Y. S. Dedkov, M. Fonin, U. Rüdiger, and C. Laubschat, "Graphene-protected iron layer on Ni(111)," *Applied Physics Letters*, vol. 93, pp. 022509–+, July 2008. 100, 101, 105, 116

[157] G. Smith and J. Anderson, "Field emission and field ion microscope studies of the epitaxial growth of nickel on tungsten," *Surface Science*, vol. 24, no. 2, pp. 459 – 483, 1971. 100

[158] E. Bauer, "Epitaxy of metals on metals," *Applied Surface Science*, vol. 11/12, pp. 479–494, 1982. 100

[159] U. Gradmann and G. Waller, "Periodic lattice distortions in epitaxial films of Fe(110) on W(110)," *Surface Science*, vol. 116, no. 3, pp. 539 – 548, 1982. 100

[160] K.-P. Kämper, W. Schmitt, G. Güntherodt, and H. Kuhlenbeck, "Thickness dependence of the electronic structure of ultrathin, epitaxial Ni(111)/W(110) layers," *Phys. Rev. B*, vol. 38, pp. 9451–9456, Nov 1988. 100

[161] F. J. Himpsel and D. E. Eastman, "Observation of a Λ_1-symmetry surface state on Ni(111)," *Phys. Rev. Lett.*, vol. 41, pp. 507–511, Aug 1978. 101

[162] S. Hüfner and G. K. Wertheim, "Core-line asymmetries in the x-ray-photoemission spectra of metals," *Phys. Rev. B*, vol. 11, pp. 678–683, Jan 1975. 103, 105

BIBLIOGRAPHY

[163] C. Guillot, Y. Ballu, J. Paigné, J. Lecante, K. P. Jain, P. Thiry, R. Pinchaux, Y. Pétroff, and L. M. Falicov, "Resonant photoemission in nickel metal," *Phys. Rev. Lett.*, vol. 39, pp. 1632–1635, Dec 1977. 105

Acknowledgements

I want to thank Karsten Horn, William Brewer, Yuriy Dedkov, Aaron Bostwick, Björn Frietsch and Thomas Braun for participating in one or the other way in my research projects.

Besides Bill, Aaron, and Yuriy, who I have mentioned previously, I want to thank Simon Wall, David M. Smith, and Stefan Behlau for proof-reading on this thesis.

I further want to thank SPECS GmbH, and specifically Sven Mähl and Oliver Schaff.

There is another group of people, who should not be forgotten: those who supported me psychologically during my hardest time. This means particularly Julia Stähler and Frauke Bierau. Moreover, Martin Norden, Clemens Taubmann, Max Klingsporn, Sebastian Heeg, Jan Harms and Johannes Großhauser did a great job in making me better after the most stressful days.

My parents always supported me, not only during the last three years. I use this opportunity to thank them for my existence and my education. And I also want to acknowledge my grandmother for her continuous financial support.

Finally, I would like to thank those members of my little trance-family that have not been mentioned previously for all the adventures. This means particularly Jan Aßmann, Lennert Hörcher, Till Naumann and Nadine, Alewhxa Gebhardt, Jolina Semmler, Julien and Jerome Hagen, Chris Behlau, Alexander Krawcyk, Ekatherina Rozenbaum, Stuart Frier, Amaru Fezer, Anant Sharma, and all related.

i want morebooks!

Buy your books fast and straightforward online - at one of world's fastest growing online book stores! Environmentally sound due to Print-on-Demand technologies.

Buy your books online at
www.get-morebooks.com

Kaufen Sie Ihre Bücher schnell und unkompliziert online – auf einer der am schnellsten wachsenden Buchhandelsplattformen weltweit! Dank Print-On-Demand umwelt- und ressourcenschonend produziert.

Bücher schneller online kaufen
www.morebooks.de

 VDM Verlagsservicegesellschaft mbH
Heinrich-Böcking-Str. 6-8   Telefon: +49 681 3720 174   info@vdm-vsg.de
D - 66121 Saarbrücken   Telefax: +49 681 3720 1749   www.vdm-vsg.de

Printed by Books on Demand GmbH, Norderstedt / Germany